# AFTER MAN
## A ZOOLOGY OF THE FUTURE
### BY DOUGAL DIXON

## INTRODUCTION
## BY DESMOND MORRIS

ST. MARTIN'S PRESS
NEW YORK

*Published in the United States of America in 1981 by*
St. Martin's Press
175 Fifth Avenue
New York
NY 10010

Library of Congress Number 81-50345
ISBN 0-312-01163-6

*Edited designed and produced by*
Harrow House Editions Limited
7a Langley Street, Covent Garden, London, WC2H 9JA

Edited by James Somerville
Designed by David Fordham

Phototypeset by Tradespools Ltd., Frome, England
Illustrations originated by Gilchrist Bros. Ltd., Leeds, England
Printed and bound by Graficromo S.A., Cordoba, Spain

# CONTENTS

FOR
GAVIN

*The author and publishers would like to express their thanks to the illustrators of this book. They are Diz Wallis (represented by Folio); John Butler and Brian McIntyre (represented by Ian Fleming and Associates Ltd); Philip Hood; Roy Woodard (represented by John Martin and Artists Ltd); and Gary Marsh*

# INTRODUCTION BY DESMOND MORRIS

As soon as I saw this book, I wished I had written it myself. It is a marvellous idea, beautifully presented. Many years ago, as a young zoologist, I started inventing imaginary creatures, drawing and painting them as an enjoyable contrast to the demands of my scientific studies. Released from the restrictions of evolution as it really is, I was able to follow my own, private evolutionary whims. I could make monsters and strange organisms, plant-growths and fabulous beasts of any colour, shape and size I liked, letting them change and develop according to my own rules, giving my imagination full rein. I called them my biomorphs and they became as real to me as the animals and plants of the natural world.

Dougal Dixon's mind has obviously been working in a similar way, although the creatures he has brought to life are very different from mine. Instead of inventing a parallel evolution, as though it were taking place on another world, he has given himself the intriguing task of contemplating a future evolution on our own planet, closely based on species that exist at present. By waving a time-wand and eliminating today's dominant species, including man, he has been able to watch, through his mind's eye, the lesser animals gradually taking over as the major occupants of the earth's surface.

Setting his scenario in the distant future, about 50 million years from now, he has given the members of his new animal kingdom time to undergo dramatic changes in structure and behaviour. But in doing this he has never allowed himself to become too outlandish in his invention. He has created his fauna of the future so painstakingly that each kind of animal teaches us an important lesson about the known processes of past evolution – about adaptation and specialization, convergence and radiation. By introducing us to fictitious examples of these factual processes, his book is not only great fun to read but also has real scientific value. The animals on these pages may be imaginary, but they illustrate vividly a whole range of important biological principles. It is this – the way in which he has perfectly balanced his vivid dreamings with a strict scientific discipline – that makes his book so successful and his animals so convincing and, incidentally, so superior to the often ridiculous monsters invented by the cheaper brands of Science Fiction.

The only danger in reading this delightful volume is that some of you may reach the point where you suddenly feel saddened by the thought that the animals meticulously depicted in it do not exist now. It would be so fascinating to be able to set off on an expedition and watch them all through a pair of binoculars, moving about on the surface of today's earth. Personally, I feel this very strongly as I turn the pages and there is probably no greater praise that I can offer the author than that....

*Desmond Morris*

# AUTHOR'S INTRODUCTION

Evolution is a process of improvement. Hence, looking at the animals and plants of today and their interactions – the delicate balance between the flora, the herbivores and the meat-eaters; the precise engineering of the load-bearing structures of the giraffe's backbone; the delicate sculpting of the monkey's foot, enabling it to grasp objects as well as to climb trees; the subtle coloration of the puff-adder's skin, hiding it completely among the dead leaves of the forest floor – and trying to project all of that into the future is a near impossibility. For how can you improve upon perfection?

One trend that is foreseeable, however, is the ruinous effect that man is having on the precise balance of nature. I have taken this not unjustifiably to an extreme, with man having extinguished the species that are already on the decline and having wreaked terrible destruction on their natural habitats before dying out himself and allowing evolution to get back to work, repairing his damage and filling in the gaps left behind. The raw materials for this reparation are the kinds of animals that do well despite, or because of, man's presence and which will outlive him – those that man regards as pests and vermin. These are more likely to survive than are the highly modified and interbred domestic animals that he develops and encourages to suit his own needs. The result is a zoology of the world set, arbitrarily, 50 million years in the future, which I have used to expound some of the basic principles of evolution and ecology. The result is speculation built on fact. What I offer is not a firm prediction – more an exploration of possibilities.

The future world is described as if by a time-traveller from today who has voyaged the world of that time and has studied its fauna. Such a traveller will have some knowledge of today's animal life and so he can describe things with reference to the types of animals that will be familiar to the reader. His report is written in the present tense as if addressed to fellow time-travellers who have voyaged to the same period and wish to explore the world for themselves.

Sit back, fellow time-travellers, and enjoy the spectacle and drama of the evolution of life on your planet.

*Dougal Dixon*

Dougal Dixon
Wareham 1981

*The sketches on this page are selected from the author's own working drawings and were used by the artists to prepare the plates and illustrations in After Man.*

# EVOLUTION

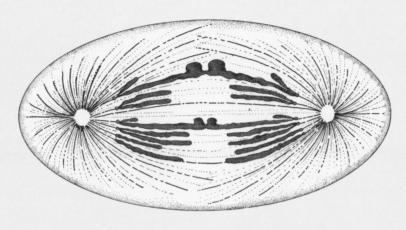

*The biological cell, shown here in the process of replicating itself, is the fundamental building block that makes up all living things. The cell's capacity for infinite variation, when taking part in sexual reproduction, is at the root of evolutionary development.*

The form and position of living things on earth can be attributed
to two things – evolution and environment. The study of evolution explores how life originated,
how it diversified the way it did and how different creatures have
developed from others. The study of a creature's environment (ecology) shows how
the various life forms interact with one another and how they interact with the environment they inhabit.
In other words evolution can be thought of as showing a longitudinal section
through the life of our planet while ecology shows the same situation in cross-section.
Each is inextricably entwined with the other and the two cannot be studied totally independently.
Although both aspects deal with survival it should not be forgotten
that extinction is a very important factor. Without it there would be no room for evolution
to take place. There would be no new ecological situations for nature to fill by the evolution of new
animals and plants from older stocks.
That evolution has taken place is apparent both from the fossil record
and from the evidence contained within living plants and animals.
Examination of fossil remains reveals a general development from the simple to the more
complex and also the part played by the environment
in shaping an organism to prevailing conditions. In living creatures,
comparability in structure, embryonic development and chemistry are powerful indications of similar
evolutionary history or of common ancestry.
Evolution is therefore not a process that has happened only in the past in order
to establish the animals and plants of today's ecology, but is a constantly continuing
process that we can study both from its results and from the fossil evidence of the past. It has happened,
it is happening now and it will continue to happen as long as life remains on this planet.

# CELL GENETICS

Animals, and indeed plants, are composed of microscopic bricks called cells. The cells found in different organs and tissues of the same creature are of quite different sizes and shapes – bones are made from angular cells, kidneys from spherical cells, nerves from long, narrow cells – but all are made from similar components. Round the outside of each cell is a skin, the cell membrane, enclosing the gelatinous cytoplasm which carries a number of small structures called organelles. The most important of these is the cell nucleus, which lies at the centre of the cell and carries the information from which the entire organism is built.

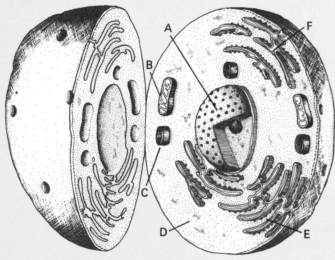

*Most animal cells contain the same basic components. At the centre lies the nucleus (A), which contains the cell's genetic material. The mitochondrion (B), responsible for energy production, and the lysosome (C), which secretes chemical products, lie nearer the surface in the cytoplasm (D). The ribosomes (E), where the proteins are assembled, lie along a convoluted structure of membranes known as the endoplasmic reticulum (F).*

This information is stored as a code, made up from a sequence of components contained in a long molecule of a complex substance known as deoxyribonucleic acid (DNA). The DNA molecule is a little bit like a ladder that has been twisted throughout its length. The shafts of the ladder are made up of sugar-phosphate molecules and each rung consists of a pair of molecules known as nucleic-acid bases. There are only four of these bases and the sequence in which they are found along the twisted ladder gives the coded instructions from which the whole organism is formed. Although repeated in its entirety in the nucleus of each cell of the organism, only certain parts of the code are needed to build up particular organs.

The peculiar thing about the DNA molecule is its ability to reproduce itself. The molecule splits along its length and unwinds so that each half of the ladder consists of a shaft and a series of half-rungs. The missing ladder halves are built from the pool of sugar-phosphate bases, which is supplied by the creature's food and is present in each cell nucleus. As each of the four types of nucleic acid base in the strand attracts only a specific kind of nucleic acid base to itself, when two new complete strands of DNA are formed they are absolutely identical to each other in the sequence of their components. This is the most important process involved in cell multiplication and underlies the growth of all organisms.

However, to grow, organisms also require proteins in the form of either structural elements such as collagen, in the case of the packing tissue between organs, or as enzymes which aid specific biological processes. Although the production of proteins is carried on outside the cell nucleus it is controlled by the DNA and is produced in a way analogous to DNA replication. The messenger that transmits the DNA's instructions to the protein production centre, the ribosome, is a molecule known as RNA. It is formed along partly "unzipped" sections of DNA and differs only subtly from it. The messenger RNA travels to the ribosome, where it links up with another form of RNA, transfer RNA, which bears amino

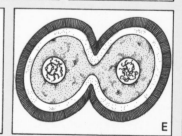

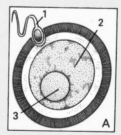

FERTILIZATION
1. Sperm
2. Ovum
3. Ovum nucleus
4. Chromatid

*The sperm penetrates the ovum (A) and comes to lie alongside the ovum nucleus (B). The chromosomes of both sperm and ovum divide into separate strands known as chromatids. Corresponding chromatids move to opposite ends of the ovum (C), where they are surrounded by nuclear membranes (D). The structure then splits into separate cells (E).*

*During cell division, when new cells are being formed, the DNA (A) contained within the dividing cell unzips and forms new molecules of DNA along its free ends (B) from the nucleic-acid bases and sugar phosphates contained in the cell nucleus. To produce messenger RNA, the DNA comes apart partially (C) and links with broadly similar material; the sugar phosphate backbone is slightly different chemically and one of the nucleic acids is substituted. The messenger RNA moves to the ribosomes, where it links up with transfer RNA, which carries amino acids (D). The messenger RNA contains the code that ensures that the transfer RNA is linked together in the correct sequence to produce the chain of amino acids that form the desired protein.*

acids. It is from these amino acids that the proteins are formed. The RNA molecules are merely code carriers and ensure that the amino acids link together in the correct sequence to form the protein type required. In this way DNA controls the workings of the whole cell and hence of the whole organism.

The DNA molecules in the cell nucleus are aggregated into structures called chromosomes, and specific groupings of nucleic-acid base sequences on the DNA give rise to specific traits in the organism. These groupings are called genes. Half the chromosomes in a creature's cells, and hence half its genes, come from its mother and half from its father. This is reflected in the alignment of the chromosomes during cell division. The chromosomes then are arranged in pairs, mother-donated ones aligned with identical father-donated ones so that comparable genes are side by side. Even though each gene in a pair contributes to the determination of a particular characteristic, one gene often masks the effect of another.

As part of the reproduction process special cells known as gametes – that is sperms or eggs – containing only half the number of chromosomes found in ordinary cells, are formed in the sex organs. Although one chromosome from each pair is present in each gamete, none is identical to any of the chromosomes received from either the mother or the father, but contains a mixture of material from both parents. This characteristic of gamete chromosomes is primarily responsible for the variation between individuals of the same species that is seen in nature. During fertilization, the gametes unite with others from a second individual to produce a complete cell, with the full number of chromosomes, which in turn divides and builds up a completely new organism with genetic characteristics derived from both parents.

This, briefly, is the sophisticated mechanism that enables plants and animals to reproduce and pass on their distinctive traits from one generation to the next. It is small changes, or mutations, in the genes involved in this process that allow evolution to take place. A mutation results in a variation in the characteristics of the adult organism growing from the cell containing the gene. In most cases the change that takes place is harmful and gives the organism a disadvantage in the competitive world outside. The organism perishes and the mutant gene perishes with it. Occasionally however the mutant gene produces a trait that gives the organism a distinct advantage in its fight for survival.

The variation in genetic make-up that sexual reproduction makes possible produces the range of characteristics that are found throughout individuals of a single species. Natural selection, which may be thought of as the directional impetus of evolution, acts on this variability, favouring certain characteristics and rejecting others according to their survival merit.

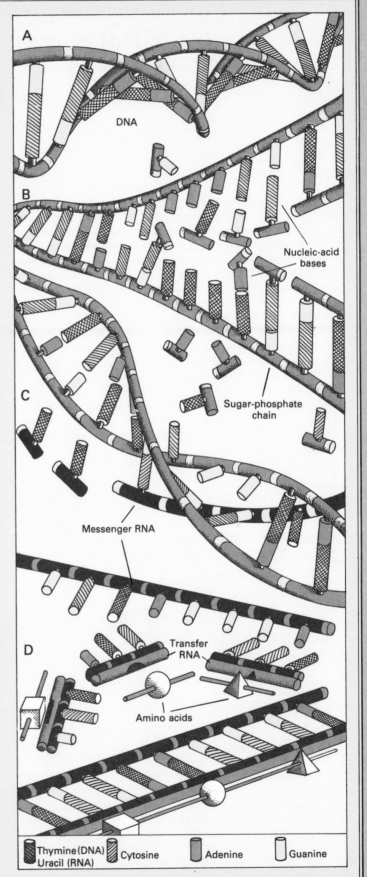

A

DNA

B

Nucleic-acid bases

Sugar-phosphate chain

C

Messenger RNA

D

Transfer RNA

Amino acids

| Thymine (DNA) Uracil (RNA) | Cytosine | Adenine | Guanine |

# NATURAL SELECTION

Natural selection, resulting from the environmental conditions in which an organism lives, can have one of three different influences on a population. It can be stabilizing, directional or diversifying. The stabilizing influence can be seen where conditions have remained unchanged over a long period of time. The resultant environment consequently supports a well-balanced population of animals and plants in which evolutionary development is disadvantageous. Under such circumstances any change occurring in a plant or animal will bring it out of the environment's neat, efficient, time-honoured survival pattern and put the creature at a disadvantage, eventually resulting in its extinction. Its more conservative contemporaries on the other hand will survive. Animals that have been subjected to stabilizing selection for a long period of time may seem quite unspecialized and primitive compared with those of similar ancestry that have experienced a more eventful evolutionary history. Often they are characterized by passive survival mechanisms such as heavy armour, or high fecundity to offset losses through predation.

The directional influence of natural selection is more evident when the environment itself changes. Under these circumstances evolutionary changes occur such as to give the impression that the organism is evolving along a set path with a particular goal in view. This is quite erroneous and arises from the fact that in the context of its environment the most recent member of an evolutionary series always appears much better adapted than the earlier intermediate stages which, where they are known, look half-formed and incomplete by comparison, even though they were equally well adapted to the environment's own earlier intermediate stages. An example of this is the evolution of the horse, which developed from a small forest-living browser into a large, long-legged running grazer as its environment altered from forest to open grassy plain. The small changes that enabled it to deal most effectively with its changing environment were continually selected for throughout its history and in this way the horse evolved.

The diversifying influence of natural selection takes effect when a new environment is established offering a fresh range of food

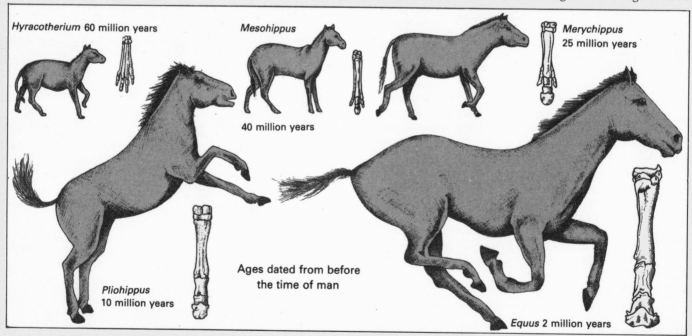

Hyracotherium 60 million years

Mesohippus

40 million years

Merychippus 25 million years

Pliohippus 10 million years

Ages dated from before the time of man

Equus 2 million years

The horse's earliest known ancestor, Hyracotherium, a small long-toed creature no bigger than a dog, inhabited the extensive forest areas found on the earth between 50 and 60 million years before the Age of Man. As conditions became drier at the end of the Tertiary and the woodland receded, the creature became progressively better adapted to life on the plains. Its feet changed radically; the outer toes disappeared, leaving a single horny hoof. Its legs became longer as it evolved into a fully fledged running animal and its dentition and digestive system changed from that of a browser to that of a grazer as its diet altered from leaves to grass. The most important structural changes occurred about the time of Merychippus, which appeared about 25 million years ago.

*At the time of man, a chain of sub-species, or cline, existed around the North Pole with the British lesser black-backed gull,* Larus fuscus graellsii, *and the British herring gull,* Larus argentatus argentatus, *as end members. All neighbouring species of the cline could interbreed with one another excepting the end members, which, by the time the chain was complete, were too distantly related to mate with one another successfully.*

(1) British lesser black-backed gull, *Larus fuscus graellsii,* (2) Scandinavian lesser black-backed gull, *Larus fuscus fuscus,* (3) Siberian vega gull, *Larus argentatus vegae,* (4) American herring gull, *Larus argentatus smithsonianus,* (5) British herring gull, *Larus argentatus argentatus*

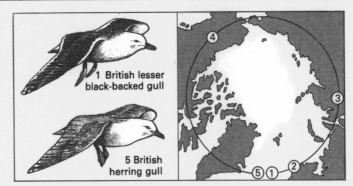

eating forms with heavy bills and a form that ate burrowing grubs winkled out with cactus spines. The large number of resulting species reflected the large number of ecological niches available on the islands.

Birds, with their power of flight, are usually the first vertebrates to reach a new island and consequently far-flung islands can usually be counted upon to produce an interesting bird fauna. Typical are the heavy flightless birds, such as the moa, *Dinornis*, of New Zealand, the dodo, *Raphus*, of Mauritius and the elephant bird, *Aepyornis*, of Malagasy, all of which evolved in the absence of ground-living predators. The intervening sea was an effective barrier preventing interbreeding between the far-travelled individuals that reached the island and the original stock back home. Such barriers to interbreeding are necessary in the evolution of new species.

Races or sub-species often co-exist in the same area, exploiting slightly different environments or food resources but retaining the ability to interbreed. They may even exist as a chain of sub-species reaching from one region to another, each sub-species able to interbreed with the next one to it. When the species at the ends of the chain are quite different the chain is called a cline. Occasionally a cline may form a ring, for example round a mountain range, where the two end members, although next to one another and related, are so different that no interbreeding is possible and are, technically speaking, different species. This poses problems in taxonomy since, as interbreeding is possible elsewhere throughout the ring, the members must strictly be considered as sub-species of the same species.

Once a group becomes isolated from its original population it may develop on its own to such an extent that, if the isolating barrier later disappears and the two populations once more intermingle, interbreeding is no longer possible. They are now, by definition, two different species. The differences are accentuated if the new location the isolated group finds itself in is basically unsuitable. The group will very quickly disappear except for maybe a few individuals at the extremes of the species range that show some slight affinity for the environment. The species that then develops will be descended from those few individuals that were genetically different from the main population in the first place and contained by chance genetic traits which made them innately more likely to survive.

Because organisms are capable of infinite variability and have an inherent tendency to change when set in an unstable environment, new species appear more rapidly when the environment is changing quickly. Evolution is so efficient that no ecological niche is left vacant for long. Something will always develop to fill it.

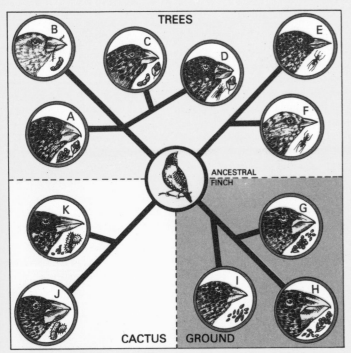

*From the original finch that arrived at the Galapagos Islands from South America, around fifteen separate species evolved to fill the island's vacant ecological niches – each species with specialized characteristics suited to its own individual diet. The finches fall broadly into three distinct groups according to habitat – cactus, tree and ground dwellers – and differ mainly in the shape of the bill. It is thought that to begin with birds were scarce on the island allowing the finches to evolve forms suitable for all the environmental slots available.*

(A) *Platyspiza crassirostris,* (B) *Cactospiza heliobates,* (C) *Carmarhynchus parvulus,* (D) *Carmarhynchus pauper,* (E) *Pinaroloxias inornata,* (F) *Certhidea olivacea,* (G) *Geospiza fortis,* (H) *Geospiza magnirostris,* (I) *Geospiza fulginosa,* (J) *Geospiza conirostris,* (K) *Geopiza scandens*

resources and living spaces. An animal species entering this environment may well evolve different forms that are specifically adapted to each of these living spaces, or ecological niches. In the absence of competing animals these different forms will eventually develop into completely new species. This is the kind of thing that happens when an island, or a group of islands, is thrown up by volcanic activity in the open ocean. The unpopulated island is slowly colonized by animals which gradually diversify into different species to exploit the whole area effectively. The classic example of evolutionary diversification is seen in the Galapagos Islands of the Pacific Ocean. Early in their history a small finch arrived that subsequently evolved into tree-living, insect-eating forms, seed-

# EVOLUTION
# ANIMAL BEHAVIOUR

Evolution does not involve the conscious will of the organism. Nor does it happen through any adaptation that is forced on it by its surroundings, or any strategy learned by the organism during its lifetime being passed on by it to its offspring. It happens, simply, because certain characteristics in an organism's genetic make-up are either selected for or selected against by the particular characteristics of environment in which it finds itself. The environment, in this context, is the physical surroundings of the organism, such as the topography, the temperature or the rainfall, and the other organisms that coexist with it, both those that it feeds on and those that feed on it.

The rate of evolution has little to do with the rate at which genetic mutation occurs – the important factor is the environment's rate of change; the speed at which new pathways open up into which new forms may evolve and develop.

As well as being responsible for structural and morphological traits in an animal the genetic make-up of a cell also gives rise to behavioural traits that allow an animal to interact with its neighbours and with its environment in a way that ensures its survival.

It can be argued that the function of an organism is merely to pass on its genes to the next generation. Evidence to support this view can be drawn from patterns of behaviour seen in animals. Behaviour is, simply, an animal's active response to its environment, and along with growth and reproduction is one of the factors that defines a living thing.

Birds crowd together when a hawk appears, thus making it more difficult for the hawk to seize an individual. Running herbivores dodge about to escape a swifter predator so that it becomes exhausted before catching any of them. Young birds stay close to

*The visual courtship display of birds is an important part of a behavioural pattern that also includes song; male bird song is designed to attract females as well as to deter rival males. Visual display may take place independently with the intention of attracting a mate. An individual, usually a male, postures and signals until it has secured the attentions of a potential mate. The pair then display in concert, each responding to the other's gestures with the object of discovering the other's willingness or readiness to mate. Many species rely on*
*resplendent plumage for display. In most cases the males are ostentatiously feathered, whereas the females are drab by comparison. The movements and gestures in courtship display are usually those associated with aggression or appeasement. In some species preening and mock sleep are all part of display.*

(A) Gannet, *Morus bassanus*, (B) Sage grouse, *Centrocercus urophasianus*, (C) Cormorant, *Phalacrocorax carbo*, (D) Brolga crane, *Grus rubicunda*, (E) Great crested grebe, *Podiceps cristatus*, (F) Adélie penguins, *Pygoscelis adeliae*

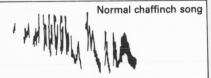

Normal chaffinch song

Song of young chaffinch reared in isolation

Chaffinch
*Fringilla coelebs*

*Investigations into the song of the chaffinch,* Fringilla coelebs, *have provided a fascinating insight into the role of learning in behaviour. It was found, as indicated on the sound spectrograph shown opposite, that young chaffinches reared in isolation were capable of only a rudimentary song and that to produce the fully developed form they had first to hear the song of others in the wild.*

their mothers until they are mature enough to fend for themselves. These, like all aspects of behaviour, have evolved to aid survival. A gene that introduces a behaviour pattern that does not contribute to the survival of the species is soon eliminated.

Courtship rituals are a very complex aspect of behaviour. The exact motion of a bird in a display dance or the movement of a lizard's head as it approaches a prospective mate indicates to its future partner that it is in breeding condition and that it is a member of the correct species. The latter point is important, for although mating between two related but separate species may produce offspring, they will almost certainly be sterile. Such matings are a total waste of time and effort from the point of view of evolution, as they do not successfully propagate the creature's genes and are therefore to be avoided.

These activities are all instinctive hereditary behaviour patterns. Other behaviour patterns are learned and are also ultimately derived from the animal's genetic make-up. The ability to establish the appropriate action by trial and error, or by the example of others around it, is an ability conferred on an animal by its genes.

Aggression is an element of behaviour that is perhaps more complicated than it first appears. One might ask why, if the object of aggression is to remove one's competitors, do not animals fight to the death each time there is a conflict? Apart from the obvious risk involved, the answer is probably that, as an animal has no chance of killing all its potential rivals, by killing an isolated one it is just as likely to assist its competitors as to benefit itself. In most cases combat in the animal world takes the form of mock battles and aggressive displays which do little physical damage to the creatures involved, but do establish the dominance of one or other of the participants. Thus the animal that wins a contest achieves what it has set out to do, that is to gain or retain the resource in dispute without suffering injury itself. The loser also derives benefit in that he escapes serious injury and retains the possibility of contesting future issues, where he may eventually be successful. It is difficult to see how this strategy could be learned and it is more likely that it is the product of evolutionary development; those animals adopting the strategy are more likely to reproduce and therefore the genes responsible for the behaviour are passed on in preference to others that result in less successful behavioural patterns.

Throughout the animal kingdom behaviour patterns are designed to ensure the survival of the individual's genes rather than the survival of the individual. Loyalty is shown to the closest relatives, since the closer the relative the larger the number of similar genes in its make-up.

The protective instinct which causes a mother bird to put itself in danger or even sacrifice its life in order to save its brood is a

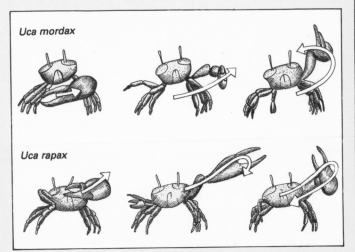

*Uca mordax*

*Uca rapax*

*Male fiddler crabs,* Uca spp., *attract mates by waving their large fiddle claws. The gestures they make both in the shape of the movement and in its speed varies between species living in the same area and ensure that only females of the correct species are attracted. As only matings between individuals of the same species are likely to produce fertile offspring, those males not possessing the genes that produce the correct waving patterns are likely to disappear.*

behavioural trait calculated to promote the survival of its own genes. As the genes of the mother bird are present in the brood and the several members of the brood have a better chance of reproducing and spreading their genes than has the single parent bird, it is to the advantage of her own genes to preserve the lives of her chicks even at the expense of her own. Less obvious is the gene-survival behaviour of social insects, such as bees and ants. A member of such a group will fight to the death indiscriminately to ensure the survival of the colony. In this case members of the colony are much more closely related to one another in genetic make-up than are other animals within a single breeding population. The survival of the colony therefore ensures the survival of the individual's genes despite the death of the individual.

Many mating ploys, particularly those seen in birds, may seem actually to reduce the individual's chance of survival rather than to increase them. The breeding plumage of many male birds, as well as being attractive to a mate, makes them visible to predators. Birds possessing particularly long and spectacular tail feathers must find them a great disadvantage when escaping from a predator. It is possible such handicaps to survival may be devices to show just how successful the male is – if it can survive with all that working against it, then it must be good! Hence the female is instinctively attracted to the male that puts on the most extravagant display.

# EVOLUTION
# FORM AND DEVELOPMENT

Natural selection lays down rules about precisely which form of life is most suitable for colonizing a particular environment. This evolutionary feature can give rise to a large number of different animals with the same superficial appearance. When the animals concerned have evolved from the same ancestor and have developed independently along similar evolutionary lines, they are said to have evolved in parallel. When the ancestors are different and the animals have evolved along quite different lines to produce the same final shape, their evolution is called convergent. An example of parallel evolution can be seen in the development of *Equus*, the horse, which appeared at the end of the Tertiary in North America, and *Thoatherium*, a remarkably similar ungulate which evolved at the same time in the then isolated continent of South America. The two forms developed independently along similar lines from similar ungulate ancestors in response to the same set of environmental

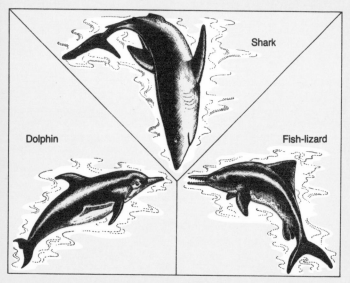

*Of the shark, fish-lizard and dolphin, only the shark has evolved from a marine creature. The fish-lizard and dolphin were evolved from a land-living reptile and mammal respectively. Despite their radically different ancestry they have all adopted the same streamlined form to suit their aquatic mode of life, and together form a striking example of convergent evolution.*

conditions. An example of convergent evolution is found in the development of the shark, *Carcharodon*, the fish-lizard, *Ichthyosaurus*, and the dolphin, *Delphinus* – three animals from totally different classes but having adopted the same streamlined shape, swimming fins and tail in order to exploit the same niche in the same environment, that of active fish-eaters in the sea.

One consequence of particular animal shapes fitting particular ecological niches is that widely separated places with the same climatic and environmental conditions may support very similar faunas even though they have evolved from different stocks. The tropical grasslands of South America, Africa and Australia all at one time supported animals with similar physical characteristics – long-legged, running grazers, swift carnivores, burrowing insectivores and slow-moving heavy browsers. In Australia they were marsupial, in Africa placental and in South America of both types. Despite their differing ancestry many of these creatures were outwardly similar. Such situations arise not only in different places at the same time, but also in different places at different times.

The influence of latitude on animal shape and form has two oddly contrasting effects. One known as Bergman's rule predicts that, within related groups, animals living nearer the poles will be larger. The other, Allen's rule, states that, again in related groups, those living nearer the poles will have smaller extremities. Both effects are heat-conservation measures designed on the one hand to preserve body temperature and on the other to prevent frostbite.

Genetic changes may be minor and quite imperceptible or they may result in changes that alter the species dramatically. The land snail, *Cepaea memoralis*, lives in a variety of habitats in the temperate woodland and can have any of several different shell markings. Where the ground is open and grassy a plain yellow coloration disguises the snail best and snails with other markings are easily seen by predators and quickly devoured. Where the ground is covered by leaf litter, brown striped forms are better camouflaged and other forms are selected against. This gives rise to populations of predominantly yellow snails in open grassy areas and brown striped snails in woodland. A similar effect was observed in the peppered moth, *Biston betularia*, during the early days of man's industrial revolution. Up until then the species had consisted largely of grey and white speckled individuals which were perfectly camouflaged against the lichen-covered tree trunks where they lived. A black form also found in the population was easily seen and eaten by birds and was therefore uncommon. With the arrival of heavy industry the trees became caked with soot and turned black, affording a perfect camouflage background for the black form. The white form was then selected against by predators and the moth population became predominantly black. Later, with the coming of clean air laws, the atmosphere and the tree trunks became less soot-laden and the moth population swung back to give a bias towards white and grey individuals once more. These changes involved only varieties within the same breeding population and there was at all times a constant exchange of genetic material as they took place. If, however, the environmental changes had been permanent and the

In looking at the life on the grasslands of Africa and Australia around the time of man and comparing it with the life that existed on the plains of South America some time earlier, during the middle Tertiary, we can see that animals with similar life styles appear to evolve similar shapes and sizes in corresponding environments. It makes no difference whether these environments are separated by time, space or both, they are by far the most important single evolutionary factor governing the shape and form of living creatures. Large herbivorous animals, very similar in appearance to the rhinoceros, and long-legged, swift-running grazing animals appeared in all three environments. Carnivores, insectivores and omnivores all superficially similar to one another evolved. The most strikingly similar groups were the burrowing insect-eaters and the flightless birds, which because of their highly specialized modes of life developed along broadly the same lines.

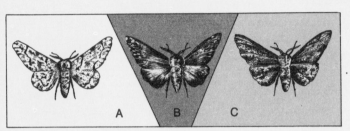

*Changes in the environment due to the industrial revolution gave the black mutant forms (B and C) of the peppered moth,* Biston betularia, *an advantage in urban areas, where they largely replaced the previously predominant grey and white speckled form (A). Because of the low level of atmospheric pollution, the population in rural areas was for the most part unaffected.*

different varieties had become isolated from one another, they would have in time become different species.

Mimicry is a separate imitative phenomenon in which a creature, usually for reasons of defence, takes on the physical appearance of another animal or of a plant or indeed of a totally inanimate object like a bird dropping. In the case of animals mimicking other animals there are two important forms. The first, known as Mullerian mimicry, occurs when a number of dangerous or unpalatable species evolve the same coloration or patterning to gain protection by association. Animals exhibiting this form tend to have vivid colours which make them stand out against the background and act as a warning, The second form, Batesian mimicry, involves totally harmless creatures adopting the coloration or appearance of inedible or dangerous species in order to take advantage of their warning coloration and so escape predation. Other forms of mimicry exist that enable predators to approach prey which they themselves mimic. The insects and in particular the butterflies with their striking wing patterns are the masters of mimicry, but it is also found among the vertebrates and among the plants.

As we have seen the rate of evolution is largely dependent on the rate of change of the environment rather than on any trait possessed by the animal itself. Even so it seems that the higher a creature is situated on the evolutionary ladder the more rapidly it evolves. For example, bivalve genera exist for, on average, about 80 million years, fish genera for 30 million years and ungulate and carnivore genera for six to eight million years. The shorter the life-span of a genus the more quickly another evolves to take its place. This results in a larger turnover of genera in land-based habitats, where life on the whole is more highly evolved than in the sea.

Plants tend to evolve much more slowly than animals and the flora existing during the Age of Man consisted mainly of plants that evolved at the beginning of the Cretaceous period while the dinosaurs were still the dominant form of land animal.

| | SOUTH AMERICA | AFRICA | AUSTRALIA |
|---|---|---|---|
| **HEAVY BROWSERS** | Astrapotherium | Rhinoceros | Diprotodon |
| **GRAZERS** | Thoatherium | Zebra | Kangaroo |
| **CARNIVORES** | Prothylacynus | Hunting dog | Tasmanian wolf |
| **SMALL OMNIVORES** | Neoreomys | Dormouse | Bandicoot |
| **ANT EATERS** | Stegotherium | Aardvark | Numbat |
| **BURROWING INSECT EATERS** | Necrolestes | Golden mole | Marsupial mole |
| **FLIGHTLESS BIRDS** | Phororhacos | Ostrich | Emu |

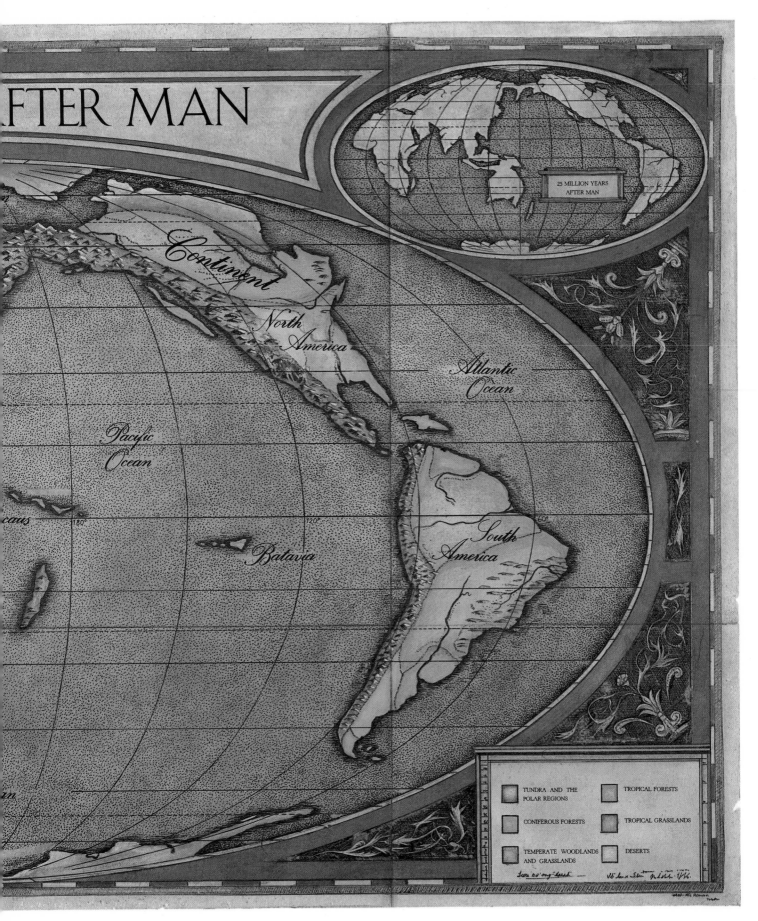

# AFTER MAN

25 MILLION YEARS
AFTER MAN

Continent

North
America

Atlantic
Ocean

Pacific
Ocean

180°

120°

60°

Batavia

South
America

| | |
|---|---|
| TUNDRA AND THE POLAR REGIONS | TROPICAL FORESTS |
| CONIFEROUS FORESTS | TROPICAL GRASSLANDS |
| TEMPERATE WOODLANDS AND GRASSLANDS | DESERTS |

# TEMPERATE WOODLANDS AND GRASSLANDS

*Across the Northern Hemisphere the temperate woodlands and grasslands form a broad belt encircling the globe, interrupted only by high mountains and seas. South of the equator temperate habitats are found only in isolated pockets.*

Temperate woodlands and grasslands are characteristic of middle latitude areas, where warm sub-tropical and cool sub-polar air masses meet. This boundary is not fixed but moves north and south with the seasons and varies a great deal according to the geography and relief of the region. In the lower temperate latitudes, the western edges of the continents tend to have hot, dry summers and mild, damp winters, while the eastern edges are warm and humid all the year round with frequent summer thunderstorms. In higher latitudes the cool sub-polar air masses are the more dominant influence and the general eastward movement of the air brings rain to the western margins, giving damp, humid conditions in both summer and winter.

The typical vegetation in humid areas is deciduous forest, but, in places where the rainfall is high and there is little difference between summer and winter temperatures, evergreen forests of both coniferous and broadleaved trees are found. Most of the tree species present are influenced by soil type and local relief. Pines are found on gravelly soils and rock outcrops, and alders and willows on waterlogged soil by rivers and streams – but the main types of tree are oak, ash, maple and beech. The characteristic feature of deciduous woodland is the difference between its summer and winter aspects. In the summer the leaves form an almost continuous canopy and little direct sunlight reaches the ground. After the annual shedding of leaves the trees stand stark and naked against the wintry skies and the inhabitants are faced with new conditions of lighting and cover as well as of temperature and precipitation.

They react to this in many ways, including hibernation and migration. The discarded leaf matter forms a thick, rich soil and contains three sources of plant nutrients – rotting plant material, humus and clay minerals. The humus slowly releases nutrients into the soil and also traps essential minerals such as nitrates and phosphates. The clay minerals store potassium, sodium and calcium – important raw materials necessary for photosynthesis.

In areas of seasonal rainfall where the total precipitation is between 25 and 75 centimetres, grass forms the dominant vegetation. Although all grassland areas have an annual period of drought lasting several weeks or months when the surface soil dries out completely, their fundamental characteristic is the total absence of moisture at depth in the soil. The lack of water at this level does not impede the growth of grass, which is shallow rooted, but prevents trees, which have deep roots, from establishing themselves.

The temperate woodlands and grasslands probably represent the habitats that suffered most during the Age of Man some 50 million years ago. Man cut down the forests to supply fuel and to provide space for agriculture and settlement. He ploughed large tracts of grassland to plant cereals and created wide expanses of pasture land for grazing animals. These disturbed areas did not revert to their natural state until a long time after man's disappearance. This interference caused the extinction of a great number of animal genera native to the original habitats. However, some creatures did survive, and it was from these that the animals of today's temperate woodland areas are descended.

A much more lightly built creature with long ears and a short sandy-coloured coat, the desert rabbuck stands no more than 1.2 metres high at the shoulder and is found in arid areas throughout the world to the south of the temperate belt.

## MOUNTAIN RABBUCK
*Ungulagus scandens*

This is the smallest and the least common of all rabbuck species and is found along the western mountains of the Northern Continent. It is adapted to live on a meagre diet of poor grasses and herbs.

## DESERT RABBUCK
*Ungulagus flavus*

## COMMON RABBUCK
*Ungulagus silvicultrix*

Heavily built with rolls of insulating fat, the Arctic rabbuck is found in the far north, in the region of the tundra and coniferous forest.

The Arctic rabbuck has a thick hairy coat which turns white in the winter.

## ARCTIC RABBUCK
*Ungulagus hirsutus*

The forest-dwelling rabbuck of temperate latitudes is the archetypal species of the genus Ungulagus. It grows to around 2 metres high and has a dappled coat which camouflages it effectively among the trees. They are normally found in small herds of between ten and twelve individuals.

# THE RABBUCKS

*The evolution of the major group of herbivorous animals*

Several species of hopping rabbuck *Macrotagus spp.* still survive. This evolutionarily older group consists largely of woodland animals that feed on the leaves and shoots of trees.

During the period immediately before and during the Age of Man the principal large-scale grazers and browsers were the ungulates, the hoofed mammals. They were generally lightly built running animals, able to escape quickly from predators and with teeth particularly suited to cropping leaves and grasses. The ungulates were widely used by man for his own purposes. Cows and goats were domesticated for milk and meat, sheep were bred for wool and the skins of many were used for leather. Horses and oxen were harnessed to work for man and became the classic beasts of burden. By the time man became extinct these animals had become so dependent on him that they could no longer survive.

The deer, the wild ungulates of the temperate latitudes, fared little better. Vast tracts of temperate woodlands had been destroyed to make room for man's cities and to provide agricultural land. This interference with their habitat was so intolerable and put such pressure on the deer that their numbers fell to a level from which they never recovered. What then could take their place? A whole ecological niche was vacant with nothing to exploit it. Which creature was best placed to take the initiative?

During the Age of Man a small-scale grazer was present that was so successful it was considered to be a pest. The rabbit was so seriously destructive of man's crops, that man made numerous attempts to control it and even attempted to exterminate it. Yet no matter what actions he took he never succeeded in getting rid of it completely. After man's disappearance, the rabbit's versatility and short breeding cycle enabled it to develop successfully into a number of separate forms. The most successful, the rabbuck, *Ungulagus* spp., now occupies the niche left by the ungulates.

To begin with the rabbuck changed little from its rabbit ancestors excepting for size. In an environment totally devoid of large, hoofed grazing animals the rabbit was left with no major grazing competitors and quickly evolved to occupy the position they once held. The early rabbucks, *Macrolagus* spp., retained the hopping gait of their forebears and developed strong hind legs for leaping. However, although jumping was ideal for moving around the open grasslands, their traditional habitat, it was not the best method for the confined spaces of the forest, and a more fundamental change had to take place. Several species of this earlier line still exist, but their place has largely been taken by the running forms of rabbuck that more closely resemble the deer of earlier times.

The second major development took place some ten million years after the Age of Man. As well as developing rapidly into the size of a deer the rabbucks also began to evolve the typical deer leg and gait. The jumping hind limbs and the generalized forelimbs of the rabbit grew into long-shanked running legs and the feet changed radically. The outer digits atrophied and the second and third toes grew into hoofs, strong enough to bear the animal's weight. This was a highly satisfactory arrangement and this line has now largely replaced the leaping form as the dominant group.

The rabbuck has been so successful that it is found in a wide variety of forms throughout the world – from the tundra and coniferous forests of the far north to the deserts and rain forests of the tropics.

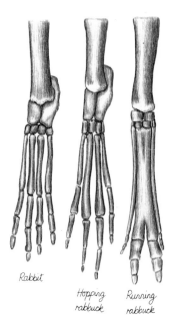

Rabbit

Hopping rabbuck

Running rabbuck

The development of the foot from the large springboard structure of the rabbit to the light, two-toed running hoof of the rabbuck was crucial to its evolution. The three principal stages are shown here, although not to scale.

The hopping rabbuck moves in a bounding motion (a,b,c,d) reminiscent of its rabbit ancestors, whereas the running form, *Ungulagus spp.*, moves in a manner similar to that of the ancient deer (e,f,g,h).

a    b

c    d

*The hopping rabbuck*

e    f

g    h

*The running rabbuck*

# THE PREDATORS

*The rise of the predator rats – the earth's principal carnivore group*

*The rapide, Amphimorphodus longipes, a native of the northern plains, is built for speed. Its highly flexible spine gives it the added impetus to reach speeds of over 100 kilometres per hour*

*The janset, Viverinus brevipes, is a long-bodied, burrowing predator, strongly resembling the extinct stoats and weasels, and like them will swim, climb trees and tunnel underground in the pursuit of its prey.*

In the mammal world the predators were traditionally carnivores (members of the order carnivora) – specialized meat-eating animals with teeth modified for stabbing, killing and tearing flesh. Their legs were designed for leaping and producing a turn of speed that could quickly bring their chosen prey within killing distance. Wolves, lions, sabre-tooths, stoats – these were the creatures that fed on the docile herbivores and kept their numbers in check both during and before the Age of Man. However, being very specialized, these species tended not to have a great life span. They were so sensitive to changes in the nature and the populations of their prey that the average life of a carnivore genus was only six and a half million years. They reached their acme just before the Age of Man, but have since decreased in importance and are now almost extinct except for a number of aberrant and specialized forms found in the coniferous forest of the far north and in the South American Island Continent.

The place of the carnivores, as the principal mammal predators, is now occupied by a variety of mammal groups in different parts of the world. In temperate regions the descendants of the rodents occupy this niche.

When the carnivores were at their peak, the rodents, particularly the rats, began to acquire a taste for meat and animal waste. The spread of man to all parts of the world encouraged their proliferation and after man's demise they continued to flourish in the refuse created by the disruption and decay of human civilization. It is this adaptability that has ensured their survival.

Despite the specialized nature of their teeth, rats were able to live on a wide range of foods. At the front of their mouths they had two sharp gnawing incisors, which continued to grow throughout life to compensate for wear and which were separated by a gap from the back teeth. These were equipped with flat surfaces for grinding vegetable matter. This is very different from the typical carnivore dentition, which had cutting incisors at the front followed by a pair of stabbing canines and a row of shearing teeth at the back.

As the rats expanded to occupy the niches left by the dwindling carnivores their teeth evolved to fulfil their new role. The gnawing incisors developed long, stabbing points and were equipped with blades that could cut into and grip their prey. The gap between the incisors and the back teeth became smaller and the grinding molars became shearing teeth that worked with a scissor action. To make the dentition effective the jaw articulation changed from a rotary grinding motion into a more powerful up-and-down action. This dentition was crucial in the development of the predator rats and allowed them to radiate into the numerous forms and varieties seen throughout the world today.

In temperate latitudes the larger herbivores, the grazers and browsers of the plains and forests that were one time prey to the wolf, have now become the prey of the falanx, *Amphimorphodus cynomorphus*, a very large dog-like rat which hunts in packs. The evolution of this form involved the modification of the limbs from the fairly generalized scampering legs of the rat to very sophisticated running organs with small, thickly padded feet, and long shanks powered by strong muscles and tendons.

*The ravene, Vulpemys ferox, is about the size of the extinct fox or wild cat and preys on small mammals and birds. It has long claws and pointed stabbing fangs.*

*The falanx is the commonest species of predator rat found in the temperate latitudes.*

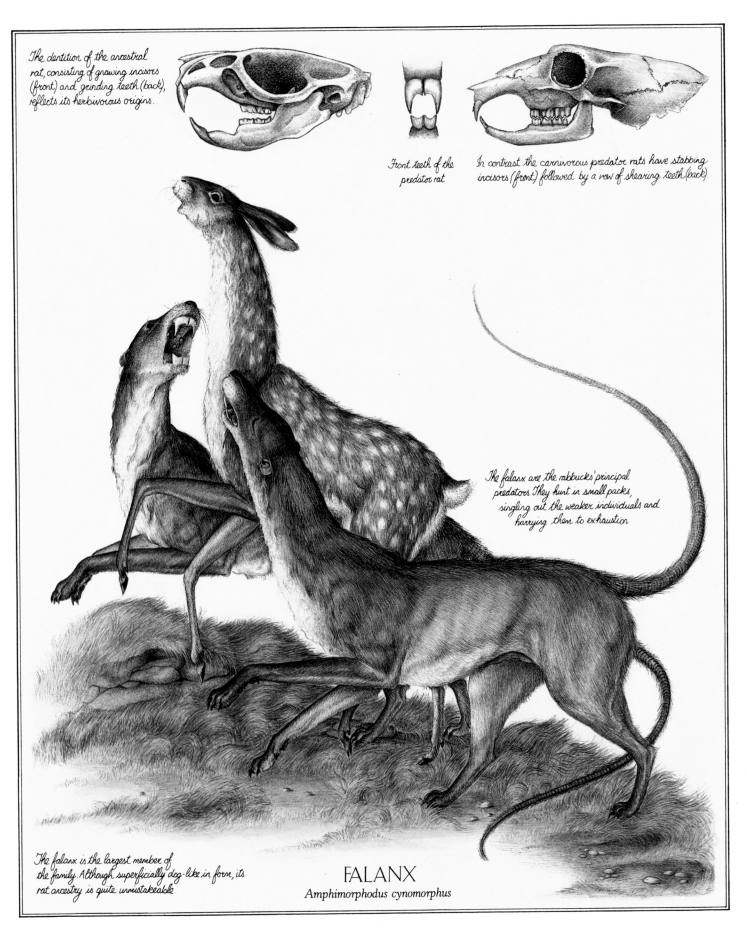

The dentition of the ancestral rat, consisting of gnawing incisors (front) and grinding teeth (back), reflects its herbivorous origins.

Front teeth of the predator rat

In contrast the carnivorous predator rats have stabbing incisors (front) followed by a row of shearing teeth (back)

The falanx are the rabbucks' principal predators They hunt in small packs, singling out the weaker individuals and harrying them to exhaustion

The falanx is the largest member of the family. Although superficially dog-like in form, its rat ancestry is quite unmistakeable.

FALANX
*Amphimorphodus cynomorphus*

# CREATURES
# OF THE UNDERGROWTH

*Life beneath the trees of the broad-leaved forests*

The tusked mole's strong limbs and powerful tusks enable it to burrow through the hardest and stoniest soils

The undergrowth of a temperate wood, thick with humus and leaf-litter and added to annually by the autumnal shedding of deciduous leaves, provides a rich source of nourishment and shelter for all sorts of animals. The primary consumers of this material are bacteria and invertebrates, such as worms and slugs, which in turn provide food for many mammals and birds. The insectivores are therefore well represented in this habitat, not only in their primitive role of small-insect eater but also in a number of varieties that have adopted a predatory, carnivorous mode of life.

Among those that have kept to their original life style is the testadon, *Armatechinos impenetrabilis*, a descendant of the primitive hedgehog. The spines of its ancestor have been replaced by a series of hinged, armoured plates which can be drawn together into an impregnable sphere when the animal is threatened. When rolled up tightly it is almost impossible to grip or penetrate and even the most determined predator rat finds a meal from this little animal more trouble than it is worth.

The tusked mole, *Scalprodens talpiforme*, comes somewhere between the old order of insectivorous animals and the newer carnivorous ones. Looking very much like a mole of 50 million years ago, it leads a burrowing existence and has adopted the streamlined shape, velvety fur and spade-like feet of its distant cousin. However, here any resemblance stops. It has two huge tusks extending from its jaws, and a paddle-shaped tail. As it burrows, the animal pushes forward with its feet in a rolling motion so that its tusks ream out the soil in front of it. The loose soil is pushed back by the feet and compacted to the tunnel walls by its tail. As well as eating worms and burrowing invertebrates, it also preys on small surface-living animals, especially mice, voles and lizards.

The most interesting example of a previously insectivorous creature turned meat eater is the oakleaf toad, *Grima frondiforme*. It gets its name from a peculiar fleshy outgrowth on its back that looks exactly like a fallen oak leaf. The toad lies partly buried in the leaf litter, totally camouflaged and quite motionless except for its round, pink tongue which protrudes and wriggles about just like an earthworm. Any small animal that approaches to investigate falls victim to the toad's powerful jaws. The animal's only real enemy is the predator rat.

These two creatures, the oakleaf toad and the predator rat, have a curious relationship. Within their blood streams lives a fluke that spends the juvenile stage in the toad and the adult stage in the predator rat. When the fluke approaches adulthood it produces a dye that turns the leaf-like outgrowth on the toad's back bright emerald green. As this happens in winter the toad becomes highly conspicuous and is quickly eaten. In this way the fluke is transferred into the body of the predator rat, where it becomes sexually mature and breeds. The fluke's eggs return to the toad through the predator rat's faeces, which are eaten by beetles that are preyed on by the toad. As the fluke needs to spend a period of at least three years growing in the toad's body before it is ready to parasitize the predator rat, and as the toad is sexually mature at eighteen months, all toads have the opportunity of reproducing before being exposed to predation.

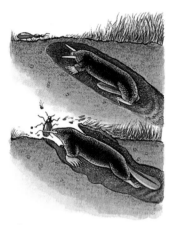

The tusked mole lies in wait just below the soil surface listening for sounds of movement above. When it hears its prey approaching, it springs out, using its tail as a lever, grasping the creature with its teeth.

The oakleaf toad lures its prey with its long worm-like tongue

When rolled-up the testadon forms
a complete sphere

The testadon has an overall length of 30 cm

Leaf outgrowth
dyed green by
nature fluke.

Untainted leaf
outgrowth
providing
camouflage.

## TESTADON
*Armatechinos impenetrabilis*

## OAKLEAF TOAD
*Grima frondiforme*

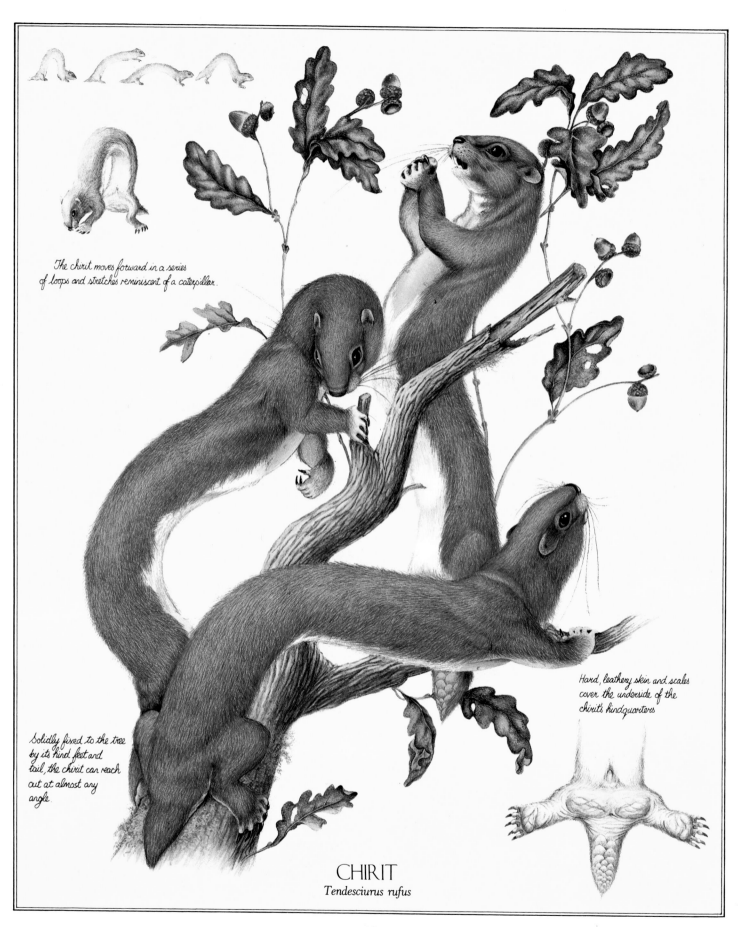

The chirit moves forward in a series of loops and stretches reminiscent of a caterpillar.

Solidly fixed to the tree by its hind feet and tail, the chirit can reach out at almost any angle.

Hard, leathery skin and scales cover the underside of the chirit's hindquarters

# CHIRIT
*Tendesciurus rufus*

# THE TREE DWELLERS

*Mammals and birds of the tree-tops*

The tree drummer's feet are covered with sensitive bristles that can detect the slightest movement in the bark beneath.

The tree goose or hanging bird in roosting position

Plant-eating mammals abound in the trees of the deciduous forests, eating shoots and leaf buds in the spring and fruits and nuts in the autumn. The long-bodied squirrel, known as the chirit, *Tendesciurus rufus*, is a typical plant-eating mammal. Its peculiar shape is a legacy from an immediate ancestor – the tree-burrowing rodent of the northern coniferous forests. As it spread south into the temperate woodlands it found that it no longer needed to make deep tunnels in the trees to escape the harsh winter, and as a result the animal's specialized chiselling and gnawing teeth became smaller, its dentition reverting to be more like that of its distant ancestor the grey squirrel. Its bodily shape, however, was still perfectly adapted to life in the trees and remained unchanged.

Now that the animal no longer led a burrowing existence, its legs and feet had to evolve to suit its new environment. Its hind feet, although small and short, became very powerful and developed strong, gripping claws. The underside of its short tail grew hard and scaly and with its hind feet formed a strong three-point anchor that could secure the animal to the tree while it reached out to collect food.

As its squirrel ancestor's jumping ability has completely disappeared, the animal can only move from one tree to another by reaching out and grasping an extended branch. For this reason the chirit is found most often in dense thickets, where the trees are close together. Its only enemies are birds of prey, and it is really only vulnerable to these when feeding in the topmost branches. It retains the predilection of the burrowing squirrel for making nests in holes in trees and often occupies holes and hollows excavated by wood-boring birds.

Wood boring is the speciality of a group of insectivores known as tree drummers, *Proboscisuncus* spp. These animals, basically shrew-like in form, subsist on a diet of grubs and insects, which they gouge out from crevices in the bark. They have masses of sensory bristles on their feet and very large ears, which help them to detect the movement of grubs burrowing in the wood. When a tree drummer finds a grub it drives its chisel-like teeth into the bark to make a hole big enough to enable it to remove the grub with its trunk-like proboscis. Sometimes the grub becomes skewered on its chisel teeth and needs to be carefully plucked off before being eaten.

It is really the birds that are the masters of the trees. After the great reptiles became extinct, over a hundred million years ago, the birds expanded into an enormous number of species. Being primarily designed for flying, birds had access to the tree-tops in a way that few other animals had, and finding that they were safer there than on the ground they soon became perfectly adapted to this new habitat. As a result many woodland birds have developed feet with curved opposable toes that are ideal for gripping branches. In one species, the tree goose or hanging bird, *Pendavis bidactylus*, these toes have been reduced to two. They are permanently curved and enable the bird to hang upside down without effort. Because of the bird's size and weight, this attitude is much easier to maintain over long periods than an upright stance, and it has taken to spending long periods roosting in this position.

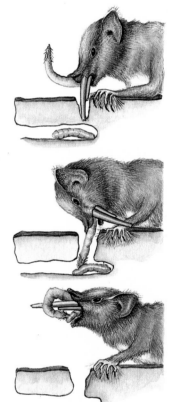

After making a hole in the bark with its chisel teeth, the tree drummer removes the grub with its gristle-tipped proboscis.

45

# NOCTURNAL ANIMALS

*The night-life of the temperate forest*

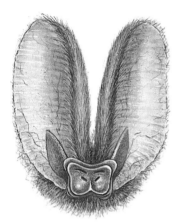

The purrip bat's sensitive ears are positioned far forward at the front of its face to provide it with the largest possible sound-collecting surface.

The shrock's outward similarity to the badger is an excellent example of convergent evolution.

The lutie's large rabbit-like ears betray its ancestry.

As night falls in the temperate woodland, the sleeping animals of the day are replaced by a completely new set of creatures. Nocturnal birds, bats and insects – a whole array of creatures is found that are as diverse and numerous as those of the daytime. As dusk falls and the moths and night-active flies take to the air the insectivorous bats appear to feed on them. Bats have proved so successful in their shape and life style that in most parts of the world they have remained remarkably stable in shape and form ever since they first appeared over a hundred million years ago. Save for the development of a more sophisticated echolocation system, positioned at the front of the face, and the absence of eyes, little else has changed.

The purrip bat, *Caecopterus* sp., so called because of its curious voice, is found throughout temperate latitudes. Unlike the earlier bats which generally navigated using high-pitched sounds, the purrip bat uses a much wider range of frequencies extending well into the audible level, giving it a much more sophisticated picture of the terrain.

Great birds of prey, which combine the characteristics of the former eagles and owls, wing their way silently through the branches, ever watching for an unwary movement on the ground that would denote the presence of a small animal. Their large forward-facing eyes, acting like wide-aperture lenses to increase the amount of light reaching the retina, give a three-dimensional image over their entire field of vision and enable them to accurately gauge distances and hunt in almost pitch darkness. Their prey includes the lutie, *Microlagus mussops*, a descendant of the rabbit.

The luties live in direct competition with the ancient groups of small rodents – the mice and voles. In some areas the luties have replaced the rodents completely, whereas in other parts of the woodlands, where the conditions particularly favour them, the rodents have remained successful. The luties resemble the small rodents in many respects, particularly in size, but their rabbit ancestry is obviously displayed in the shape of the head and tail. They feed mostly at night, nesting during the day in crevices among tree roots or in holes in the ground.

Another small animal that provides food for birds of prey is the truteal, *Terebradens tubauris*, an insectivore related to the chisel-toothed shrews of the trees. The incisors of both the upper and lower jaws of this animal are extended forward to form a structure like a bird's beak, which acts as a probe to catch worms and burrowing insects in soft earth and leaf litter. The truteal is completely blind and retains no vestiges of eyes. It is, however, equipped with a large number of sensory whiskers and extremely acute hearing. Its ears, which are enormous for the size of its body, can be rolled into trumpets by a unique set of muscles located at their base and then pressed to the ground to listen for sounds of burrowing.

The shrock, *Melesuncus sylvatius*, is a much larger animal. Descended from insectivore stock, it has a size and shape comparable to that of the extinct badger. It makes nightly forays through the undergrowth and will take any prey that it chances upon. It has a long snout and broad forepaws with which it digs after burrowing animals and excavates its own family nest in soft soil under tree roots.

The largest of the owl-eyed predatory birds stands more than a metre high.

The purrip bat's sonar equipment is typical of most temperate woodland bats.

PURRIP BAT
Caecopterus sp.

TRUTEAL
Terebradens tubauris

The truteal has an overall length of about 12 centimetres

The truteal uses its long beak to probe deep into the earth for worms and grubs.

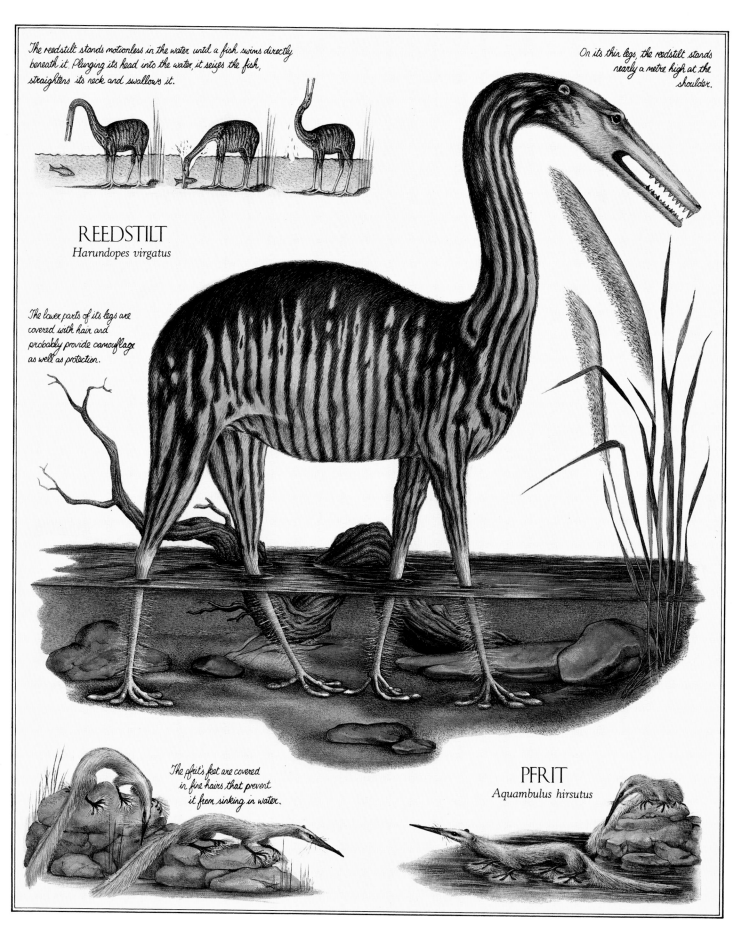

The reedstilt stands motionless in the water until a fish swims directly beneath it. Plunging its head into the water, it seizes the fish, straightens its neck and swallows it.

On its thin legs, the reedstilt stands nearly a metre high at the shoulder.

## REEDSTILT
*Harundopes virgatus*

The lower parts of its legs are covered with hair and probably provide camouflage as well as protection.

The pfrit's feet are covered in fine hairs that prevent it from sinking in water.

## PFRIT
*Aquambulus hirsutus*

# THE WETLANDS

*Life in the fens and marshes*

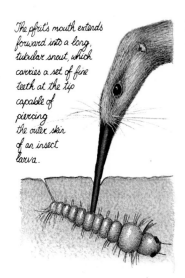

The pfrit's mouth extends forward into a long, tubular snout, which carries a set of fine teeth at the tip capable of piercing the outer skin of an insect larva.

In temperate latitudes wetland areas are comparatively isolated pockets of land found scattered widely across the Northern Continent. As well as strictly water habitats such as ponds, lakes and rivers, they also include stretches of saltmarsh and fenland found near the sea, mires and peat bogs found in poorly drained inland regions and areas of regular inundation.

The conditions found throughout this range of habitats is so diverse in terms of salinity, oxygenation, light penetration and water currents that very nearly every individual location has its own little ecosystem and associated fauna, and almost every animal group is represented.

One of the most unusual water-living mammals is the pfrit, *Aquambulus hirsutus*, a tiny insectivore descended from the primitive shrews. Its length, excluding its tail, is less than five centimetres, which puts it among the smallest mammals in existence. Although its body is thin, its feet and tail are broad and are covered with water-repellent hairs, which spread its weight over such a large area that it can skate across the water without breaking the surface tension. It lives mainly on the larvae of mosquitoes and midges that are found just under the water surface. It feeds on them by piercing their outer cuticles with its long, hairless snout and draining them of their vital juices while they are still in the water. In this way the pfrit avoids disturbing the water surface, which would both upset the surface tension and frighten away its prey.

A mammal frequently found near river banks and lake sides is the reedstilt, *Harundopes virgatus*. Its long, slender legs and neck and vertical stripes render it almost totally invisible among reeds, where it is frequently found fishing. Its head and neck are most unusual. Practically all mammals have seven neck vertebrae, but the reedstilt has fifteen. In evolutionary terms the extra vertebrae have appeared quite recently and result from the fact that, in fishing, longer-necked individuals have an advantage over the others. The tooth pattern is degenerate – the incisors, canines and molars having all reverted to an almost reptilian condition in which they are all of the same shape. The reedstilt uses this combination of neck and tooth features to catch fish by darting out its long neck and snapping shut its needle-pointed teeth.

Fishing skills have also been developed to a high degree by the angler heron, *Butorides piscatorius*. This bird, an inhabitant of the North American subcontinent, creates shallow ponds at the water's edge in the shade of overhanging trees by scraping at the river bottom and constructing shallow dams. On the shore nearby it accumulates a heap of droppings and fish remains to attract beetles and flies. These it then picks up and drops into the shallow water to entice the fish into its pond, where they are easily caught.

Although there are many examples of flightless birds, the long-necked dipper, *Apterocinclus longinuchus*, a river bird of the European sub-continent, is the only bird that spends part of its life with the capacity for flight and the rest flightless. During its early life the bird develops wings in the normal way, but once it has migrated away from its natal nesting site it becomes totally earth-bound and pursues a purely terrestrial – aquatic existence. Its wings now no longer necessary, lose their power and gradually atrophy.

The long-necked dipper

Flying wing of juvenile

The wing of the breeding adult is degenerate and used only for balance and in swimming under water.

Once the angler heron has baited its "fish pond" it remains close by, watching motionless from the reeds.

# CONIFEROUS FORESTS

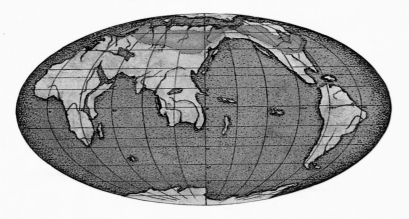

*Throughout the world coniferous forests are found in areas having the lowest temperatures permissible for the growth of trees. The largest expanses are found at the far north of the Northern Continent, bordering the tundra.*

The coniferous forests of the Northern Continent represent the greatest expanse of uninterrupted forest in the world. Coniferous trees do well at high latitudes because they are evergreen, and photosynthesis can take place immediately conditions are right for growth without having first to produce leaves, as is the case with deciduous trees. In this way the conifers compensate for the short growing season, which is about 50 to 80 days depending on the latitude. Fruiting and reproduction are also in tune with the climate. Conifers, unlike deciduous trees, do not produce fruiting bodies that are pollinated and ripen within a single year. The fertilization of a female cone may take more than a year to complete, and as many as three more years may elapse before the cone matures and the seeds are ripe for dispersal.

The lack of leaf litter and the prevailing cold conditions which inhibit the natural decay of the forest's pine needle carpet – material that is slow to decompose in any case – results in only a thin underlying layer of topsoil and little or no undergrowth. The indigenous mammals are largely herbivorous and exist mainly on diets comprising mosses, pine needles, bark and seed cones. Insectivorous birds are rare compared with those that feed on cone-seeds and buds.

Throughout the region forest fires are not uncommon, usually occurring in spring, when the trees are low in sap. Large areas can be devastated at a time. Recolonization is firstly by deciduous trees such as birches, alders and rowans that are only later replaced by the climatic vegetation of spruce, larch, cedar or pine.

The coniferous tree's typical tall, pointed shape is ideal for bearing the weight of the winter snowfall and allows the snow to be shed quickly when it melts in spring. Their surface-spreading root systems are perfectly adapted to the shallow soils that are characteristic of the habitat.

In the north of the region, where the underlying soil is frozen all year and is therefore impervious to water, there are many lakes, streams and bogs with their own localized flora of mosses and sedges. The forest is more open and blends into the neighbouring tundra. Larger patches of tundra mosses and lichens appear on high ground. Near rivers in this transitional area the forest remains thick and extends far northwards along sheltered valleys into the tundra. At the southern edge of the coniferous belt, the conifers grade imperceptibly into deciduous woodland.

Throughout the world, smaller areas of coniferous forest are found outside their normal latitudinal extent, particularly on the slopes of mountains, where the altitude produces climatic conditions similar to those experienced near the poles.

During the Age of Man the coniferous forests experienced considerable environmental damage, due mainly to clearance for agriculture and also in the course of commercial forestry. This effectively exposed large areas of land to the erosional effects of wind and rain, destroying the soil structure and consequently reducing its water-retention capacity. The coniferous forests took some time to recover from this damage, for the normal successive recolonization could not take place immediately.

# THE BROWSING MAMMALS

*The evolution of the hornheads*

The ancestral hornhead, Protocornudens, which existed between 35 million and 40 million years ago was smaller and more antelope-like in appearance. The horny head plate did not appear until sometime later.

The ornate horn structure plays an important part both in courtship and in male dominance struggles.

The browsers are the largest animals living in the coniferous forest regions. They feed mainly on young twigs and needles in the summer and subsist on bark, mosses and lichens during the rest of the year.

Across the northern continent the most prolific species are those that are derived from the gigantelopes of the African sub-continent. These northern animals, although much heavier than their distant antelope ancestors, are still not nearly as huge as the African gigantelopes. Only the shaggy tundra-dwelling forms of the far north can compare in size with these.

This difference in size between the two different northern forms is due to two separate periods of migration. The first took place about forty million years ago, before the great mountain barriers between Africa and Europe were thrown up and at about the time that the rabbuck was ousting the antelope from its traditional home on the African plains. The gigantelopes, then at an early stage of evolution, were forced to spread northwards into the coniferous forest, where they later flourished and developed into the hornheads, *Cornudens* spp.

The second migration took place more recently, about ten million years ago, when the African gigantelopes had reached their present elephantine proportions. The erosion of the mountain chain that once separated the Indian sub-continent from the rest of Asia opened up new paths to the north and led to their gradual colonization of the Tundra and the evolution of the woolly gigantelopes, *Megalodorcas* sp.

Soon after their arrival in the coniferous forest the ancestral hornheads' jaws and horns began to evolve in response to their new environment. In common with all the now almost extinct ruminants, most of these creatures possessed no upper incisor teeth. They cropped grass by working their lower incisors against a bony pad on the roof of the mouth. However, this system is not particularly effective for browsing from forest trees. The first change .that took place was that the horny head plate became extended forward to form a sort of beak. The lower lip became muscular and grew forward to meet it, thus extending the mouth some distance beyond the front teeth. This fairly primitive arrangement is still found in several species, for example the helmeted hornhead, *Cornudens horridus*. In more advanced forms, however, the lower jaw is also extended so that the lower front teeth meet the horny beak instead. These adaptations are the result of evolutionary pressure that enabled only those forms that could feed successfully on the twigs, bark and lichens of the coniferous trees to survive. The elaborate horn formation above the eyes is also used for defence.

The horn structure has been carried one stage further in the water hornhead, *Cornudens rastrostrius*, that inhabits lakesides and the banks of rivers. In this creature the horny plate extends forward into a broad rake-like structure, with which the animal grazes on soft water weeds that it finds on the beds of ponds and streams. It has two broad hooves on each foot, set widely apart and connected by a web of skin, which prevents the animal from sinking into soft mud and sand. The water hornhead, in its way of life, must surely resemble the hadrosaurs – the duckbilled dinosaurs of the latter part of the Age of Reptiles.

6 months

9 months

1 year

2 years

3 years

The hornhead's horn formation grows gradually throughout adolescence and early adulthood and in the case of the helmeted hornhead takes its final form around three years of age.

The later hornheads, such as the common hornhead, have longer lower jaws and bite by bringing the lower incisor teeth into contact with the upper horn lip.

# HELMETED HORNHEAD
*Cornudens horridus*

Like its antelope ancestors, the helmeted hornhead – the most primitive surviving hornhead – bites by bringing its lower incisors into contact with a bony pad in the upper jaw.

# COMMON HORNHEAD
*Cornudens vulgaris*

Hornheads stand about 2 metres high at the shoulder. Their horns, originally developed for a defensive role, have evolved since as browsing tools.

The water hornhead's jaws, having become extended and broadened, are used for feeding on water plants.

# WATER HORNHEAD
*Cornudens rastrostrius*

# THE HUNTERS
# AND THE HUNTED
*The relationship between predator and prey*

*Parops lepidorostrus*

*Although closely related to the broadbeak it more nearly resembles their common ancestor, the starling.*

*The broadbeak is the most massive predatory bird found in the coniferous forests.*

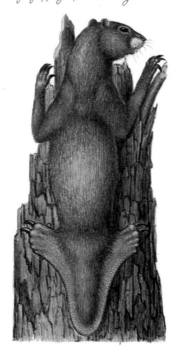

*The combined rear limbs and tail of the modern beaver form an ideal structure for gripping a pine tree's rough bark.*

As in all other habitats the animals of the coniferous forest fall into the familiar food-chain pattern of predator and prey with the carnivorous animals forming the final link. Here, as in the temperate woodlands, the fiercest and commonest hunters are the predator rats. They hunt beneath the trees in small packs, tracking down the rabbuck and the hornhead, singling out the weak and elderly individuals and running them to exhaustion. The predator rats take it in turn to attack, savaging their prey with powerful front teeth. Hornheads have such powerful horns that, when they are the quarry, it is almost as dangerous for the hunter as it is for the hunted.

A predator unique to the coniferous forest is the pamthret, *Vulpemustela acer*, a large weasel-like creature and one of the few true carnivores still in existence. Its size – up to two metres in length – makes it by far the largest predatory animal found in the region and it probably owes its survival to its low, powerful build and its ability to run through the sparse undergrowth easily, bursting out suddenly upon its prey. Pamthrets live in small family groups and normally hunt in pairs.

Not all the predators are mammals; birds also kill their share of the small animal population. The broadbeak, *Pseudofraga* sp., one of the larger birds of prey, has a wing span of over a metre and lives in the western forests of the Northern Continent. It is descended from the starlings, which expanded to fill the gap left when many of the ancient predatory birds became extinct during the Age of Man. It has a rounded tail and broad, blunt wings, which enable it to fly swiftly and manoeuvre in the tight spaces between the trees. It has a straight, powerful bill and strong talons, which it uses to grip its prey. The broadbeak's closest living relative, *Parops lepidorostrus*, is a totally different creature. It is only ten centimetres long and lives mainly on insects that it extracts from the bark of trees with its thin beak.

With so many predators in the coniferous forest it is not surprising that the smaller mammals should have evolved such a variety of defensive ploys to ensure their survival. The spine-tailed squirrel, *Humisciurus spinacaudatus*, is an excellent example of their ingenuity. It has a long, broad, flat tail with quills developed on its underside, which when at rest lie flat over the ground. However, when the animal is alarmed it throws its tail over its back and the sudden increase in skin tension erects the quills. This presents an almost impenetrable barrier and can be turned to deflect an attack from either side.

One large rodent that became adapted to a semi-aquatic way of life during the Age of Mammals, partly as a defence against predators, was the beaver. After man the beaver, *Castor* spp., became even better adapted to life in water. Its tail and hind feet have become fused together into one large paddle, which, when powered by its backbone, produces a powerful up-and-down swimming stroke. Its ears, eyes and nose are placed high up on its head and remain above water when the rest of the animal is submerged. Surprisingly the paddle does not impair the creature's movement on land and is used as a grasping limb, enabling it to climb partway up trees, increasing its potential supply of food and building materials.

*The modern beaver's swimming stroke involves the whole body.*

*Its family lodge, complete with underwater entrance, is built of sticks plastered with mud.*

The spine-tailed squirrel places a barrier of quills between itself and the carnivorous pamthret.

PAMTHRET
*Vulpemustela acer*

SPINE-TAILED SQUIRREL
*Humisciurus spinacaudatus*

# COMMON PINE CHUCK
*Paraloxus targa*

## TREVEL
*Scandemys longicaudata*

The male uses its heavy beak like a nutcracker to crush pine cones.

The sexes of the common pine chuck are so different in appearance that they look like different species.

## CHISELHEAD
*Tenebra vermiforme*

The chiselhead gnaws through the living wood to make a home for itself deep inside the tree.

56

# TREE LIFE

*Birds and mammals that feed on and in the trees of the coniferous forests*

Throughout the Age of Mammals the rodents were one of the most successful animal groups in the coniferous forests. Their powerful teeth enabled them to cope with the tough vegetable matter found there and their warm, furry coats helped them to retain body heat during hibernation.

The chiselhead, *Tenebra vermiforme*, a rodent and a relation of the temperate woodland chirits, is highly adapted to life in the coniferous forest. Its huge incisor teeth and wormlike body enable it to burrow deep into the living wood, where it can remain protected from the cold in winter. Although in some ways the animal is at an advanced stage of development, its parasitic way of life is really quite primitive. Its staple diet is the bark of trees, which it strips off completely, leaving the tree totally denuded. This, combined with the massive damage it does by burrowing, kills the tree within a few years.

As the chiselheads only colonize live trees they must be continually on the move and every spring, after hibernation, the young of the new generation migrate to find new territories. During migration they are very vulnerable and many are taken by predators before they can complete the journey. The balance between burrower and predator is highly critical and it needs only a slight reduction in the number of predators to produce an increase in the population of burrowers that would lead to the total destruction of vast areas of coniferous forest.

No other small rodent found in the coniferous forest is quite so destructive. Most live on shoots, bark and the seeds found in cones. Many are ground dwellers and feed from the cones where they fall. Others are lightly built and agile enough to scramble along the branches to where the cones are actually growing.

One large vole-like rodent, the trevel, *Scandemys longicaudata*, is unusual in having a prehensile tail. Too heavy to reach the cones growing on the slenderest branches, it feeds on them instead by hanging by its tail from a sturdy neighbouring branch and reaching out with its front paws. Like other rodents of this general size it gathers more than is necessary for its immediate needs and stores the rest for the lean winter months. Its hibernation nest is a long, drooping structure woven together from grass, strips of bark and pine needles. Built hanging from the end of a branch it is large enough to accommodate the animal together with sufficient food to see it through the winter.

Of the many seed-eating birds found in the coniferous forest, the largest by far is the common pine chuck, *Paraloxus targa*. The two sexes of this species are quite different, both in appearance and in their mode of life. The male is much more powerfully built and is equipped with a massive beak, which it uses for breaking open pine cones to feed on the seeds. The female, much smaller and drabber, totally lacks the male's heavy beak and is really a scavenger supplementing her diet with carrion, insects, grubs and birds' eggs. Most probably the common pine chuck's ancestor was a bird similar in appearance to the present-day female and the male has evolved its own distinctive features primarily for display and its eating habits are a secondary development.

Each spring the chiselheads select a new tree and construct a labyrinth of tunnels and nesting chambers in preparation for the winter.

The chiselhead's deeply-rooted incisor teeth form effective burrowing tools.

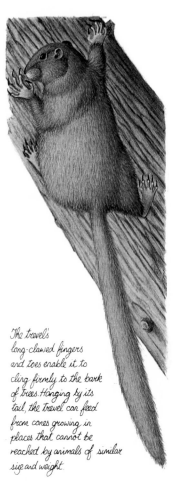

The trevel's long-clawed fingers and toes enable it to cling firmly to the bark of trees. Hanging by its tail, the trevel can feed from cones growing in places that cannot be reached by animals of similar size and weight.

# TUNDRA AND THE POLAR REGIONS

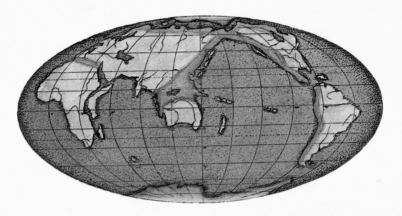

*Tundra and Arctic habitats are found at both polar extremities of the globe and at the tops of high mountains. Conditions in these localities are broadly similar and the habitats differ only in that one is an effect of latitude and the other of altitude.*

The bleakest places on the surface of the earth are found around the North and South Poles – regions of constant ice and snow where no plants grow. Because of the tilt of the earth, at certain times of the year no sunlight whatsoever reaches these regions and night lasts for months at a time. Even during summer, when daylight is continuous, the sun's rays hit the ground at such a shallow angle that very little warmth is felt. These conditions prevail both on the Southern Continent of Antarctica and on the ice-mass that covers the northern Polar Ocean.

The extent of the ice on the Polar Ocean is dependent on the low salinity of the Arctic waters – a saltier sea would not freeze over to such a degree. The Polar Ocean is separated from the Atlantic Ocean by a barrier of islands that inhibits their intercirculation. This island chain is formed from what was once a single island known as Iceland. It consisted of lavas that erupted from the mid-Atlantic ridge as the crustal plates of Europe and North America moved away from one another. As this movement continued, enlarging the Atlantic Ocean, Iceland, straddling the mid-oceanic ridge, split into two parts, each moving in opposite directions. The continuing volcanic activity spawned a string of new islands in the growing gap between the two parts. Almost 180° away, at the opposite side of the Arctic Ocean, the same crustal movements were responsible for closing the Bering Strait, the gap between North America and Asia, and fusing the two areas into one vast Northern Continent. As a result the Polar Ocean is now practically landlocked, and is fed by the rivers of the surrounding supercontinent.

The generally low-lying areas fringing the polar ice sheets comprise the tundra. During winter they are as cold and bleak as the Arctic wastes, but in summer the temperature rises above freezing and may reach an average daily temperature of 10°C. In summer the snow melts, but because of the permafrost – the layer of perpetually frozen soil beneath the surface – the water cannot drain away and floods the numerous hollows and depressions.

Spring on the tundra is a time of spectacular change. A sudden bloom of vegetation bursts forth to take advantage of the brief growing season. Much of the vegetation reproduces asexually rather than by producing seeds, as is the case in warmer regions. Vegetative reproduction is much faster and therefore, because of the short summers, much more successful. Those plants that do reproduce sexually produce seeds that are highly resistant to frost. Mosses, lichens and low bushy herbs are typical of tundra plants.

The tundra vegetation's sudden summer flourish is accompanied by a bloom of insects and in spring a veritable plague of flying creatures emerges to take advantage of the short period of warmth and sunlight. The seasonality of plant and insect life on the tundra means that for most mammals and birds food is only available during part of the year and most of the larger animals are consequently migrant, spending the harsh winters to the south.

In the Southern Hemisphere there are no equivalent large landmasses at latitudes that would produce a covering of tundra vegetation. The tundra that exists is found scattered on the islands of the Southern Ocean and on mountains just below the snow line.

# THE MIGRANTS
*The wandering herds and their predators*

The pilofile's bristles extend its potential insect-catching area beyond the region of its head.

When its beak is closed the bristles drop down, allowing it uninterrupted forward vision.

In comparison with other parts of the world, the animal and plant life of the tundra consists of a rather small number of species, each of which contains a relatively large number of individuals – a situation which is diametrically opposite to that found in the tropics. The low species count is entirely due to the region's inhospitable conditions. All tundra animals have evolved from creatures found in more temperate areas; their ancestors probably colonized the tundra only because they were driven to do so by fierce territorial competition. Life has to be unusually unpleasant elsewhere for a group of animals to venture into the tundra in the first place.

During the summer months the tundra is literally infested with flying insects and has a large population of insect-eating birds. Many, like the pilofile, *Phalorus phalorus*, have bristled beaks – a ring of stiff hair-like feathers surrounding the beak that forms a cone and deflects insects into its mouth. The bristles in effect widen the bird's potential capture area and increase its food supply.

For many large animals the tundra is only habitable during the summer months and in winter they migrate southwards into the coniferous forests, where conditions are less austere. The largest of these animals is the woolly gigantelope, *Megalodorcas borealis*, a close relation of the tropical gigantelope. It differs mainly in size and in the possession of a large, fatty hump, which provides it with nourishment during the hungry winter months. It has a long, shaggy winter coat and broad hooves, which prevent it from sinking into soft snow. It uses its enormous horns as snow ploughs to expose the mosses, lichens and herbaceous plants on which it feeds. Its eyes are small to avoid being frost-bitten and its nostrils are bordered by blood vessels that warm the air before it reaches the lungs.

In early summer, the woolly gigantelope loses its shaggy coat and takes on a much sleeker appearance. The hump which sustained it through the winter months is now entirely depleted and it spends much of the time eating to rebuild its energy store for the long trek back south in the autumn.

Because of the woolly gigantelope's size – three metres at the shoulder without the hump – there are very few predators powerful enough to threaten it. Its only real enemy, the bardelot, *Smilomys atrox*, is a creature that would have been very much at home back in the first half of the Age of Mammals. At that time elephants, animals of comparable size to the gigantelopes, were preyed on by sabre tooths. These creatures, members of the cat family, had long, stabbing canine teeth with which they inflicted deep, stabbing wounds on their quarry. After an attack the sabre tooths would wait until the elephants bled to death before moving in to feed. This successful arrangement was even evolved independently among the marsupials. However, during the Age of Man the elephants declined and the sabre tooths, being entirely dependent on them, died out completely.

With the advent of the gigantelopes the sabre tooth pattern reappeared, but this time among the predator rats. The bardelot, unlike other members of the group, exhibits sexual dimorphism in that only the female is equipped with sabre teeth and hunts the gigantelopes. The male, having none, resembles more the polar bears that once inhabited these latitudes.

In summer the pilofile feeds on the wing.

In winter it migrates south, shedding its bristles and growing in their place a long, probing bill.

The pilofile's green-and-brown-blotched egg is perfectly camouflaged in the tundra vegetation.

## BARDELOT
*Smilomys atrox*

Male bardelot

Female bardelot

Pouch for sabre teeth

The female's sabres are formed from the outer crowns of its two front teeth – a pattern derived from that of the predator rat.

Woolly gigantelope with winter coat and fatty hump.

## WOOLLY GIGANTELOPE
*Megalodorcas borealis*

The large forward-pointing horns are used to clear snow from the vegetation in winter.

Woolly gigantelope with summer coat — scruffy patches of winter coat remain throughout most of the summer.

## GANDIMOT
*Bustivapus septentreonalis*

One of the largest birds of the tundra the bootie bird stands over a metre high and is shown here in its winter plumage.

The meaching's fortress is built of vegetable matter — principally mosses and lichens. The interior is divided into chambers, one for each adult.

## MEACHING
*Nixocricetus lemmomorphus*

## BOOTIE BIRD
*Corvardea niger*

The bootie bird, so called because of the shaggy feathers it grows to protect its legs in winter, is the meaching's principal avian predator.

# THE MEACHING AND ITS ENEMIES

*A compact ecosystem*

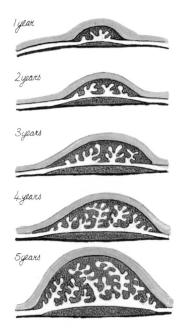

*1 year*

*2 years*

*3 years*

*4 years*

*5 years*

*Initially a four- or five-chambered nest the meachings' fortress grows exponentially, reaching its maximum size after about five years*

*The lesser ptarmigan nests exclusively in meaching fortresses.*

As the constantly frozen ground of the tundra makes digging through the earth impossible, all small burrowing rodents found in the tundra are snow-tunnellers. One tunneller in particular, the meaching, *Nixocricetus lemmomorphus*, similar to the ancient lemming from which it may be descended, has a very considerable effect on the ecology of the area.

A colony of meachings may be started by as few as three or four individuals. They breed profusely, and as their numbers grow they build a fortress of matted vegetable material to protect them from the frosts and snows. The interior of the fortress is very complex and consists of a network of passages and tiny chambers – one for each individual. During the winter each rodent is fully insulated and kept warm by the rest of the colony.

As the population of the fortress increases each year, so does the local population of predators. The meaching's principal predator is the polar ravene, *Vulpemys albulus*, a beast about the size of the extinct fox and very different from its temperate woodland cousin, *V. ferox*. It has a small head with tiny eyes and ears (an adaptation that prevents frostbite) and long, dull brown fur that turns white in the winter to camouflage it against the snow. It attacks the meaching by digging into the fortress with its front paws.

The meaching's other enemies are mainly birds. The largest is the bootie bird, *Corvardea niger*, a descendant of the crow. It has a long neck and bill and long legs, and in this respect looks rather like a heron. Indeed, in summer it even behaves like a heron, wading into shallow pools and streams to dip for fish. In the winter it develops insulating feathers along its legs to protect them from the cold and becomes a land predator, hunting any small animals that are active at the time. It probes for the meachings through the snow and, with its long beak, is able to penetrate deep inside their fortress.

The other notable avian predator of the meaching, the gandimot, *Bustivapus septentreonalis*, is descended from the magpie. It retains much of its original body shape and coloration, but has a hooked beak and pointed wings like a skua. In summer it feeds on rodents and small birds in the tundra, but spends the winters in the coniferous forests to the south as a scavenger. Its survival in the cold north is due in no small part to the fact that it is a brood parasite, laying its egg in the nests of other birds to be incubated and hatched by them. In this way it conserves the energy it would otherwise use in nest-building and brood rearing, at the expense, however, of the ducks and waders on whose nests the eggs are laid.

Even though the meachings have many predators their birth rate is so high that under normal conditions the colonies thrive, Eventually, after about four or five years of continual growth, the local food supply of herbs, seeds, mosses and lichens becomes depleted and can no longer support the colony. At this time the meachings migrate, and unprotected by their fortress fall easy prey to their predators. Up to forty per cent of the migratory population may be wiped out before finding a new habitat.

The old fortresses provide homes for several of the tundra's inhabitants. The lesser ptarmigan, *Lagopa minutus*, nests exclusively in old meaching burrows and is sometimes found cohabiting with the meachings themselves, usually in cases where part of the population has already migrated.

*Winter coat*

*Summer coat*

*In early autumn the polar ravene moults its dull brown summer coat and grows a thicker creamy-brown covering of fur*

*Although the polar ravene is larger than the ravene of the temperate woodlands, it has smaller facial features*

*Polar ravene*

*Temperate ravene*

# THE POLAR OCEAN

*Life in the northern seas*

Height 60 cm    Height 45 cm

The flightless auks exist as a chain of subspecies around the Polar Ocean capable of breeding with their neighbours excepting at the ends of the chain, where differences in size and physiology make it impossible.

The northern polar sea is almost landlocked and contains a permanent ice-cap, which has a considerable influence on the environment of the surrounding continent and contributes substantially to the stability of the region's cold climate. The ice-cap is maintained only because the Polar Ocean is fed by enormous quantities of fresh water by the rivers of the surrounding continent. This gives the sea an unusually low salinity and therefore a strong tendency to freeze over.

In winter the Polar Ocean is largely barren. In spring, however, the sunlight produces a bloom of unicellular algae near the surface, which provides food for the microscopic animal life that forms the basis of the oceanic food chain. In spring, shoals of pelagic fish come northwards through the northern island barrier to feed on the zooplankton, bringing with them countless numbers of seabirds.

The first species to arrive is the flightless auk, *Nataralces maritimus*, a totally aquatic creature with paddlelike wings. In this respect they resemble the penguins, which were so successful in the southern oceans in earlier times. Except during winter the flightless auks rarely come ashore or climb on to the ice, where they are quite defenceless. They retain their eggs until they are almost ready to hatch and lay them in the open water.

The flightless auks first evolved at the northernmost tip of the Northern Continent and, as they became established, spread both east and west, forming a chain of subspecies in a ring around the Polar Ocean. Throughout most of the ring each subspecies is able to breed with the neighbouring ones, but where the ends of the chain overlap the differences are so great that no interbreeding is possible and these populations must be regarded as separate species.

Preying on the flightless auks, and also on the fish, are the pytherons, *Thalassomus piscivorus*, a group of aquatic carnivorous mammals related to predator rats. They occupy the same ecological niche that the seals occupied earlier in the Age of Mammals and like them have developed streamlined blubbery bodies and fin-shaped limbs.

Among the organic detritus on shallower areas of the ocean bed are found banks of shellfish. Living on these shellfish is the distarterops, *Scinderedens solungulus*, by far the most massive aquatic relative of the predator rats. It reaches a length of about four metres and has an insulating coat of matted hair made up of a mosaic of solid plates, giving it a lumpy rather than streamlined appearance.

Its most unusual feature is its teeth; the upper incisors form long, pointed tusks – the left-hand one projects forward, whereas the right-hand one points straight down and is used as a pick for removing shells from the sea bottom. This asymmetry is also found in the limbs; the left foreflipper only is equipped with a strong claw, which it uses to dislodge particularly stubborn shells. Because the distarterop's evolutionary line separated from the predator rat's when they were both still comparatively small rodent-like creatures, it would appear therefore that the predator rat's double-pointed incisor teeth, from which the distarterops tusks (and also the bardelot's sabre teeth) have evolved, were a comparatively early development.

The pytheron, although totally unrelated, has an appearance similar to the ancient seals and sea-lions.

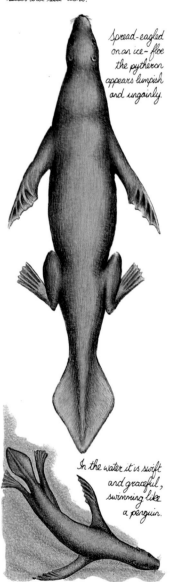

Spread-eagled on an ice-floe the pytheron appears lumpish and ungainly.

In the water it is swift and graceful, swimming like a penguin.

64

Distarterops are gregarious creatures and are often found resting on ice-floes during the summer months in small groups containing both males and females.

The females' tusks both curve downwards, whereas the males' tusks point in different directions.

## DISTARTEROPS
*Scinderedens solungulus*

In all cases the male's forward-pointing tusk is formed by the left-hand upper incisor and similarly only its left-hand flipper is endowed with a claw.

## FLIGHTLESS AUK
*Nataralces maritimus*

65

At over 12 metres long, the vortex is the largest creature on earth. It is closely related to the fish-eating porpins, which are descended from the penguins

The vortex is a plankton-eater. Its massive bird's beak has evolved into an effective sieve.

# VORTEX
*Balenornis vivipera*

# PORPIN
*Stenavis piscivora*

# THE SOUTHERN OCEAN

*The origins and ancestry of the vortex*

Skerns are found mainly around the volcanic islands of the Southern Ocean.

They have oily-green plumage, large feet and legs, but no wings.

Skerns cannot walk but use their legs to push themselves along on their bellies.

Swimming at the surface, they sit very low in the water

When hunting fish underwater they become graceful and agile swimmers.

In contrast with the vast Southern Continent, which supports life only around the edges, the surrounding ocean teems with life. Among its most notable inhabitants is the vortex, *Balenornis vivipera*, the largest animal found anywhere in the world. Resembling many of the sea creatures of the past, it has a long, tapering, neckless body, a powerful paddle-shaped tail and long stabilizing fins – an ideal arrangement for efficient movement through water. Similar shapes can be seen in the great arthrodires of the Age of Fishes, in the pliosaurs of the Age of Reptiles and in the whales of the first half of the Age of Mammals – the last creatures to occupy this ecological niche before the vortex.

The vortex is in fact descended from the penguins, which, although they were birds, had long since lost the power of flight and were totally adapted to an aquatic life excepting for one thing – they always had to come on shore to lay eggs. This remained so until, shortly after the extinction of the whales, one species of penguin developed the ability to retain its single egg internally until it was ready to hatch and gave birth to live young in the open ocean. Freed of the necessity to come ashore, this species became completely marine and ultimately gave rise to a completely new order of marine birds, the Pelagornids, of which the porpin, *Stenavis piscivora*, is the commonest surviving example.

The Pelagornids are unique in the aquatic world in that, like their ancestors, they are both warm-blooded and egglayers, albeit that their eggs are retained within the body until the moment of hatching. In this respect they resemble the mammals and some reptiles. However, it is important to note that Pelagornids do not possess mammary glands with which to feed their young, as do mammals, and are warm-blooded as reptiles are not.

The porpin, like most of its class, is a fish-eater. Its distinguishing feature is a long, serrated beak that enables it to catch larger fish than would otherwise be possible. So successful has it been that it has remained virtually unchanged for the last 40 million years.

Although a plankton-eater and very much larger, the vortex is also a member of the order Pelagornid. Its beak has developed into a large plankton sieve, which consists of a very fine mesh of bone plates instead of coalesced hair, as in the case of the whales' baleen plates.

Around the volcanic islands of the Southern Ocean are found the skern, a species of flightless seabirds that have evolved a unique behavioural quirk in response to the problem of incubating eggs in this hostile environment. As well as being a hazardous time for the embryo chicks the parent birds also run the risk of exposure. The skern has solved the problem by laying its eggs in the warm volcanic sands of the islands and deserting them immediately afterwards. It is able to delay the time of laying until the temperature of the sand is exactly right. When a volcano shows signs of activity it immediately becomes the scene of frenzied activity. The birds scramble ashore, and with the aid of their temperature-sensitive beaks probe the sand for areas with the right condition for incubation. After laying their eggs ten to twenty centimetres deep and covering them with sand they return to the sea, seeing neither their eggs nor their offspring again.

# THE MOUNTAINS

*The effect of altitude on animal communities*

The ruffle's front teeth are designed for eating mosses and lichen.

Shaggy hair on the undersides of its legs and on its feet give it a booted appearance.

The ruffle is sure-footed over boulders and loose scree.

The flora of the mountains has much in common with that of tundra regions because of the similarity in climatic conditions found there – low temperature, high precipitation and short growing season in both habitats.

Although the mountain areas of the world are so isolated and widely distributed that they can largely be regarded as separate faunal provinces, the fauna of the fold-mountain belt between Africa and Europe show characteristics that are typical of mountain life the world over. The ruffle, *Rupesaltor villupes*, a descendant of the rabbit, exhibits many of these features. It has a rounded head and body, and disc-like ears – adaptions that guard it against cold. It has long hair under the neck and body to protect its legs from the cold and its teeth are well adapted for grazing mosses and lichens. The upper incisors are set at an angle and are used for scraping the patchy vegetation from the surfaces of rocks and boulders.

The groath, *Hebecephalus montanus*, a variety of small hornhead frequently found grazing on grassy, south-facing slopes, lives in small herds of four or five females, guarded jealously by a male. The most apparent difference between males and females is in their horn structure. The males have flat, bony plate-like horns which they use to buffet one another in their frequent fights for herd dominance. The females' pointed pyramidal horns are much more deadly and are used to defend themselves and their young against predators. While the herd grazes, the male normally stands on a promontory watching for signs of danger. When it sees an intruder the male signals by erecting its long flag-like tail and the herd makes for the shelter of a nearby crag or cave.

One of the deadliest predators found in the African-European mountains is the shurrack, *Oromustela altifera*, a carnivore related to the weasel-like pamthret, *Vulpemustela*, of the northern coniferous forests. Sure-footed over difficult rocky terrain and well camouflaged by its mottled grey fur it is the groath's principal enemy. The shurracks hunt in packs, surrounding their prey, or cornering them in ravines, sharing the kill among themselves.

Perhaps the strangest mammal found in these regions is the parashrew, *Pennatacaudus volitarius*. The adults are unremarkable small shrew-like creatures, but the juveniles possess one of the strangest devices found in the animal kingdom. At the end of their tails, they have a fantastic parachute structure formed of interwoven hair, which they normally use only once before discarding. When the time comes to leave the parental nest, they launch themselves into the air, relying on the thermal currents that rise from these bare rocky slopes in summer to carry them to a fresh habitat, in some cases several kilometres away. As a means of dispersal this is a bit hit-and-miss, but the inevitable high death rate that this behaviour produces among young parashrews is more than compensated for by the large numbers of offspring produced by each adult breeding pair.

The evolution of the parashrew's parachute tail is primarily due to the creature's insectivorous ancestry. It is thought that these early creatures used their tails as balancing organs when leaping to catch insects in mid-air. The parachute consists of soft, curled hairs hooked together to form a mat and held in shape by a series of bristles growing from the tip of the tail.

The young parashrew's single migration flight may last for up to 24 hours.

The parachute tail is present only during adolescence and is moulted when the parashrew becomes sexually mature.

The animal life of mountainous areas is found at high altitude
only during the summer months. In winter, when the snows come,
herbivores such as the groaths move to lower levels to find shelter
and grazing. The carnivorous animals, dependent on them for
food, closely follow the herds on their seasonal migrations.

The male groath
stands lookout
on a prominent
rock, alert
for signs
of danger.

## GROATH
*Hebecephalus montanus*

Although female groaths are well
equipped to defend themselves,
they will not normally fight
unless forced to do so.

The adult shurrack's coat is mottled
with dark rosette markings.

## SHURRACK
*Oromustela altifera*

# DESERTS:
# THE ARID LANDS

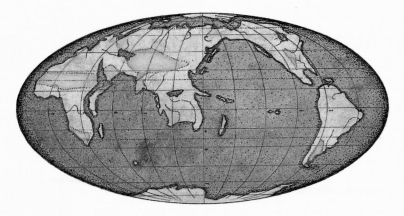

*The world's hot deserts lying along the tropics are a product of the earth's atmospheric circulation. The cold deserts occurring in the Northern Hemisphere owe their origin more to their position in the centre of large land masses.*

The principal desert areas of the world are found in the two subtropical belts between latitudes 10° and 35° north and south of the equator. Very little land in the Southern Hemisphere lies between these latitudes, with the exception of the southernmost tip of the African sub-continent and the narrow tail of the South American island continent, and therefore the major deserts of the world lie in the Northern Hemisphere. Desert conditions are characterized by extreme dryness; annual precipitation is less than 25 centimetres, and the sun evaporates all water that falls as rain.

The lack of moisture reaching desert areas is due to a number of factors. The principal one is the descent of dry air from the upper atmosphere, a feature of the global circulation pattern that is typical of these latitudes. Air is drawn to the low-pressure zones in equatorial regions, where it is heated and forced to rise. In the upper reaches of the atmosphere the air spreads out from the equator, cools gradually and descends, reaching the ground in desert areas, where it has a very low water content. In some cases a desert may owe its existence to being situated in the heart of a continent, far from any marine influence or moist wind. Allied to this is the rain-shadow phenomenon, found in areas where air currents from the sea cross high mountains, dropping their rain on the seaward side as they rise. On the other side of the mountain the air descends, completely dry, giving rise to a desert area.

The intensity of solar radiation in the desert is very great in comparison with other areas. In moist regions up to 60 per cent may be reflected away by clouds, atmospheric dust, water and plant surfaces, but in deserts only ten per cent of the total radiation is reflected in this way. On the other hand, due to the lack of insulating cloud cover, up to 90 per cent of the accumulated heat of the day is lost at night through radiation. The result is an extremely large temperature difference between day and night, which may be as great as 40°C.

The worldwide extent of deserts is now much smaller than was the case during the Age of Man. For one thing there is a smaller area of land lying within the desert belts than there was then – due largely to the movement of the Australian continent northwards out of the desert belt. Also in man's time, inefficient agricultural techniques and widespread grazing of domestic animals on poor land artificially enlarged the desert regions and was one of the factors that led to man's ultimate decline; with the desert becoming more extensive year by year, the area of the earth's surface suitable for cultivation decreased. After man's disappearance the earth's natural habitats re-established themselves and the deserts resumed their natural proportions.

Life in the desert has to cope with many hostile factors, such as lack of water and extremes of temperature. Even so a large number of animals and plants have evolved to cope with them successfully. Similar compensating adaptations – such as highly efficient kidneys producing particularly concentrated urine, large ears to dissipate heat and the ability to burrow to depths at which conditions are less severe – have been developed independently in many widely separated groups of animals in different areas of desert.

# THE SAND DWELLERS

*Survival in a waterless sea*

The leaping devil is armed with long talons and a mouthful of sharp teeth.

When pouncing on its prey, the leaping devil may travel up to 2 metres in a single jump.

The physiology of desert animals must tread a narrow path between water conservation and heat regulation. The lack of sweat glands – a water-saving measure – means that less conventional ways must be used to cool the animal in the heat of the day. Usually this is achieved by large ears or similar outgrowths which, criss-crossed by blood vessels, act as radiators to remove the animal's body heat.

A structure of this kind is found on the tail of the sand flapjack, *Platycaudatus structor*, a fairly large rodent found in sandy areas. Its excess body heat is carried away by the blood to the tail, where it is dissipated into the atmosphere. When pursued the animal can move at speed, running with its lengthy tail held well out behind as a counterbalance in the manner of its ancestor, the jerboa.

To conserve water the flapjack even constructs a condensate trap. As part of their courtship ritual each pair of flapjacks places a pile of stones over the site of the family burrow. These stones as well as protecting the burrow from the sun's direct rays during the day provide a large number of cold surfaces on which moisture can condense at night.

The spitting featherfoot, *Pennapus saltans*, is a rodent of a type that has existed in this environment ever since mammals first colonized the hot, dry areas of the earth's surface. It has small forelimbs and long hind legs, for jumping. The toes are fringed by short, stiff hairs. Its kidneys are highly efficient, recycling the animal's waste water to the point that its urine is more than twice as concentrated as that of a rodent of similar size living in a humid environment.

The featherfoot never drinks water but obtains all the moisture it needs from plants. It can even eat plants that are poisonous to other animals, and has the ability to excrete toxic substances without their having taken part in any metabolic process. It is a nocturnal animal, but if displaced from its deep burrow by a predator during the heat of the day, it can cool itself by producing copious quantities of saliva and coating the front of its body with foam. It also spits at its foe with deadly accuracy. As the saliva contains most of the excreted poisons from the plants it is an efficient weapon. Needless to say this defence mechanism dehydrates the animal very quickly and is therefore used only in dire emergency and even then only for short periods at any one time.

The leaping devil, *Daemonops rotundus*, an insectivore with carnivorous habits, is one of the featherfoot's chief predators and has a morphology and physiology similar to the featherfoot and to the other small desert rodents on which it preys.

A totally different predator, but one also descended from insectivore stock, is the desert shark, *Psammonarus* spp. It is sausage-shaped with a blunt, strong head and powerful shovel-like feet. It swims through the sand rather than burrowing, bursting into the rodents' nesting chambers, which it locates using the sensory pits at the end of its nose. It is almost completely hairless and avoids the extremes of temperature by remaining underground for most of the time. When it is at rest it lies just below the surface with only its eyes and nostrils protruding.

The flapjacks co-operate in moving stones to construct their condensation traps.

By foaming, the spitting featherfoot both cools itself and gets rid of unwanted toxins.

Although it is held out behind when running, the flapjack holds its tail aloft to catch the cooling breeze when standing still.

72

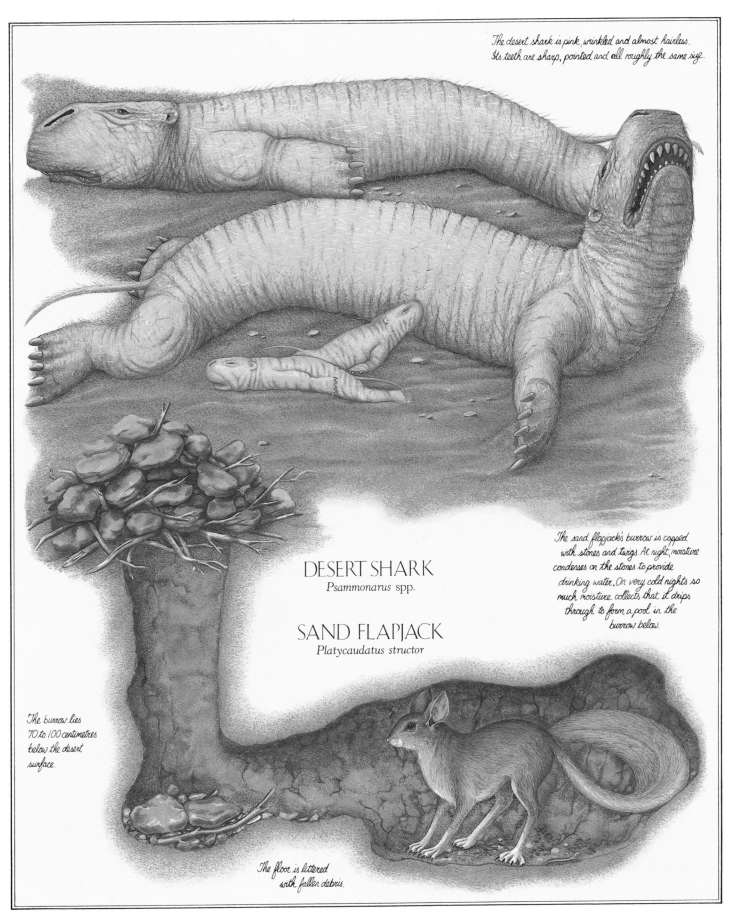

The desert shark is pink, wrinkled and almost hairless.
Its teeth are sharp, pointed and all roughly the same size.

# DESERT SHARK
*Psammonarus* spp.

# SAND FLAPJACK
*Platycaudatus structor*

The sand flapjack's burrow is capped
with stones and twigs. At night, moisture
condenses on the stones to provide
drinking water. On very cold nights so
much moisture collects that it drips
through to form a pool in the
burrow below.

The burrow lies
70 to 100 centimetres
below the desert
surface.

The floor is littered
with fallen debris.

73

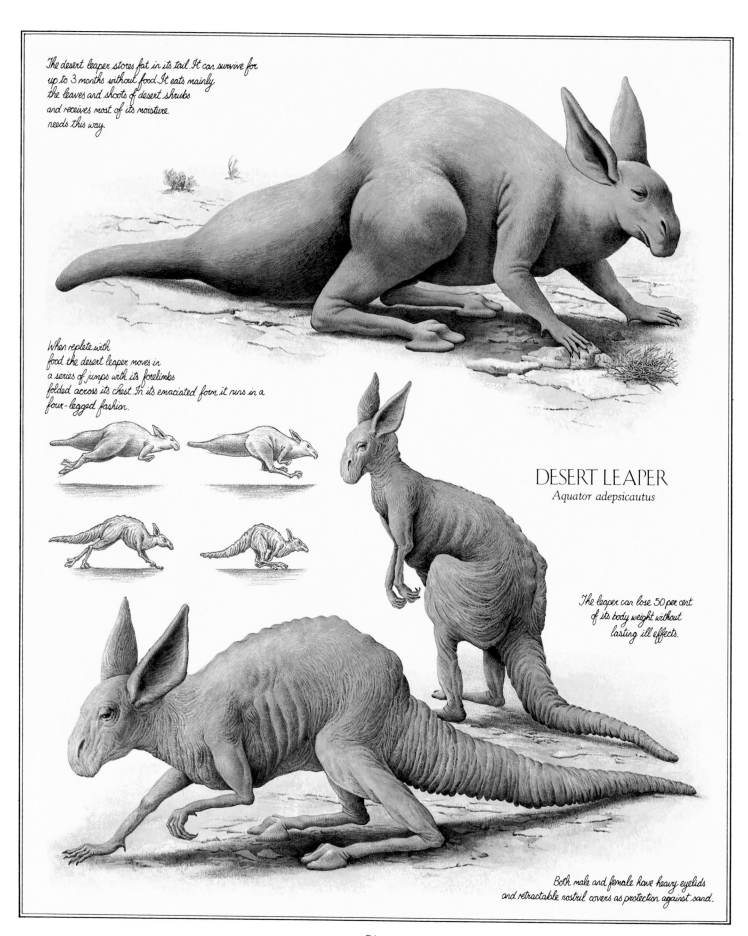

The desert leaper stores fat in its tail. It can survive for up to 3 months without food. It eats mainly the leaves and shoots of desert shrubs and receives most of its moisture needs this way.

When replete with food the desert leaper moves in a series of jumps with its forelimbs folded across its chest. In its emaciated form it runs in a four-legged fashion.

DESERT LEAPER

*Aquator adepsicautus*

The leaper can lose 50 per cent of its body weight without lasting ill effects.

Both male and female have heavy eyelids and retractable nostril covers as protection against sand.

# LARGE DESERT ANIMALS

*The problem of size and its solution*

The grobbit is purely herbivorous. Its long tail may grow to over 100 centimetres.

The grobbit feeds mainly on the leaves and shoots of desert shrubs.

Its cloven hoofs and dew claws are used in conjunction to grasp branches.

The extinction of the camel at about the same time as man died out left a niche that was distinctly unattractive to any other animal. For a large animal to exist in desert conditions a quite remarkable physiology is required. The camel, for instance, was able to lose about 30 per cent of its body weight through dehydration without ill-effects, and it stored all the subcutaneous fat of its body in one lump, leaving the rest of the body free to radiate heat. It could tolerate fluctuations in its body temperature to some extent and had thick nostril covers and eyelids that effectively kept dust and sand out of its nose and eyes.

After some 50 million years of evolution these features have all developed again in one animal – the desert leaper, *Aquator adepsicautus*. The leaper is descended from the rodents, possibly one of the jerboas or sand rats, and has grown large – adult males may reach more than 3 metres from nose to tail. The tail is the most unusual feature of this animal; it is here that all its subcutaneous fat is stored. The fat is not a water store, but a store of food that enables the leaper to go for long periods without eating when food is unavailable. When the fat store is full the animal's body is well balanced and it can leap quickly along on its hind limbs. In this condition it can undertake journeys of 100 kilometres or more between waterholes and oases. It has broad, horny pads on the toes of its hind feet which prevent it from sinking into the sand and give it a good grip on naked rock.

The rocky areas of the desert are the habitat preferred by the grobbit, *Ungulamys cerviforme*. This rodent is about 60 centimetres long, excluding the tail, and has hooves developed on its third and fourth digits enabling it to run about the craggy landscape of the rocky desert. The second and fifth digits of its front feet have small claws that almost touch the hooves when the foot is bent, allowing the grobbit to grasp and pull down branches and feed on them. The grobbit lives in packs all over the rocky desert zone of the African and Asian sub-continents.

In the deserts, large predators are not common and very few meat-eating mammals of any description are found. The khilla, *Carnosuncus pilopodus*, however, descended from the insectivores, is one of the few. Standing about 60 cm high at the shoulder, it is largely nocturnal and spends most of the day hidden in a network of burrows excavated in soft sand. At night it hunts small mammals and obtains most of the water it needs from the moisture contained in their flesh.

Most desert animals are sandy yellow in colour to blend in with the surroundings and have white undersurfaces that counteract the effects of shade and give them a two-dimensional appearance. That this colour scheme is the result of evolutionary pressure is a belief supported by the darker appearance of animals found on black-grey lava areas and the almost white forms of the same animals found in salt-pan regions.

Animals that are not camouflaged are predominantly black. Predatory birds, reptiles and the most poisonous and unpalatable arthropods fall into this category. The colour resemblance may be due to a form of mimicry in which for some reason black is an advantageous colour for certain predators, and all others adopt the same colour to derive some similar benefit.

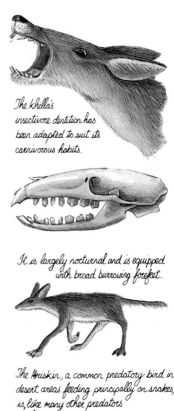

The khilla's insectivore dentition has been adapted to suit its carnivorous habits.

It is largely nocturnal and is equipped with broad burrowing forefeet.

The khiskin, a common predatory bird in desert areas feeding principally on snakes, is, like many other predators of the region, mainly black in colour.

# THE NORTH AMERICAN DESERTS

*Living in the shadow of the mountains*

*To retain as much water as possible, the rootsucker lies on the desert surface with its headshield drawn in tight against its body shell.*

The North American deserts are rain-shadow deserts. The wet westerly winds that blow towards the continent across the Pacific Ocean first meet the great western mountain barrier and are forced to rise, dropping their rain on the seaward side. Past the peaks the winds are dry and desert conditions prevail on the extensive plains beyond.

The deserts are not completely barren, but contain an intermittent vegetation consisting of cacti and other succulent plants, normally growing as single specimens, each widely separated from one another. The barren soil surface between the plants conceals a vast network of roots spreading out to collect enough water for each plant to survive.

Among the roots lives the rootsucker, *Palatops* spp., an animal heavily armoured to protect it from desiccation rather than to defend it from attack. Its head is shielded by a broad spade-like plate and its back is covered by a shiny nut-like shell composed of compacted hair. Its tail and feet are also armoured, but with articulated plates that permit total mobility. The rootsucker moves through the sand using its broad feet like paddles and its head shield as a shovel to reach the roots of succulents on which it feeds, gnawing them with the edge of its head shield and lower incisors.

Among the thorns found in the vertical grooves of cactus stems lives the little desert spickle, *Fistulostium setosum*, its narrow body covered by spines that are partly for defence and partly for camouflage among the cactus thorns. It has no teeth and subsists entirely on the nectar of cactus flowers which it drinks through its long snout. When collecting nectar it often picks up pollen on its head. The pollen is eventually deposited on the stigmas of other flowers, thus effecting the cross-pollination of the cacti. Living almost solely on nectar, the spickle's digestive system is a very primitive affair, since nectar is very easily broken down.

Lizards and other reptiles do not have the sophisticated mechanisms that mammals and birds have for regulating body temperature. Their temperature is entirely dependent on the surroundings. Several desert reptiles have, however, developed rudimentary devices for keeping themselves cool. The fin lizard, *Velusarus bipod*, for instance, a small bipedal reptile, has a system of erectile fins and dewlaps on its neck and tail which it raises into the wind when its body becomes too hot. The heat is transferred through the fins via the blood stream into the air. When cooling itself, the lizard typically balances on one leg while keeping the other off the hot desert surface to get maximum benefit from the system.

Small mammals of the desert, like the desert spickle and the fin lizard, are preyed on by ground-dwelling birds such as the long-legged quail, *Deserta catholica*. Its eggs, which are laid in sand scrapes in sheltered spots beneath bushes or overhanging rocks, are sat on continuously to protect them from the extremes of heat and cold that are typical of the desert climate's daily temperature range.

The breeding cycle of this and many other desert birds is dependent on the rainy season, the birds nesting as soon as the first spring rains appear and continuing as long as the wet season lasts. In unusually dry years no breeding takes place.

*The desert spickle sucks nectar from cactus flowers through its long snout. For such an ungainly little animal it can run surprisingly swiftly over the desert.*

*Only the male long-legged quail has a head plume.*

*The females are otherwise identical although a little smaller.*

*The eggs of the long-legged quail are laid in sheltered hollows in the desert sand.*

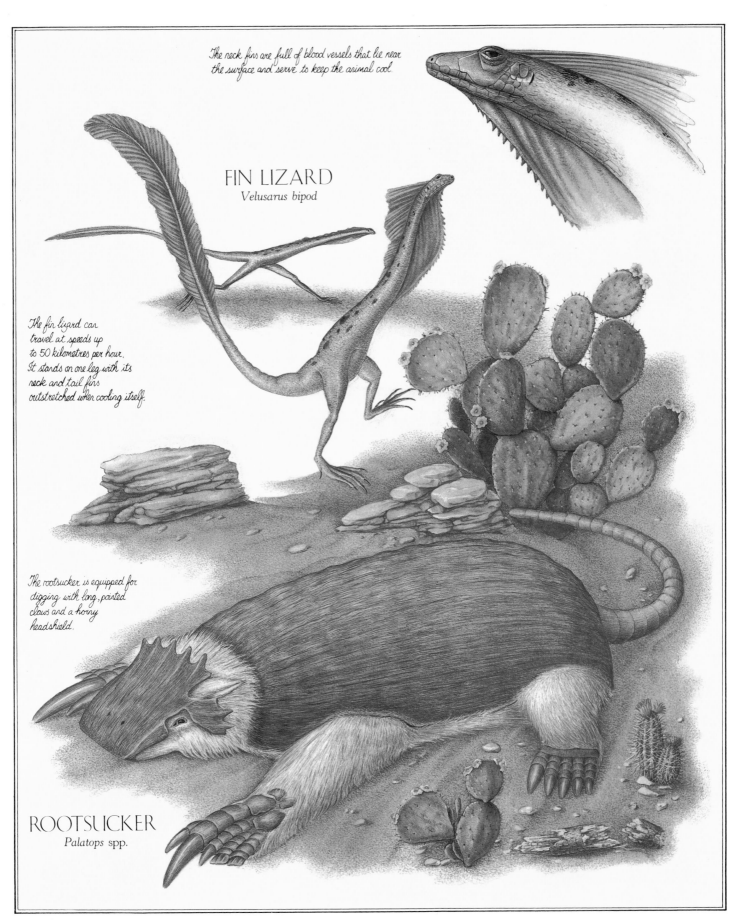

The neck fins are full of blood vessels that lie near
the surface and serve to keep the animal cool.

## FIN LIZARD
*Velusarus bipod*

The fin lizard can
travel at speeds up
to 50 kilometres per hour.
It stands on one leg with its
neck and tail fins
outstretched when cooling itself.

The rootsucker is equipped for
digging with long, pointed
claws and a horny
headshield.

## ROOTSUCKER
*Palatops* spp.

# TROPICAL GRASSLANDS

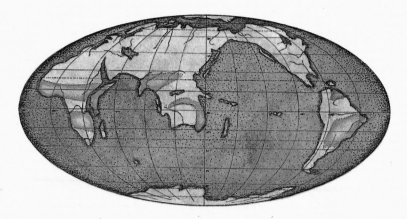

*In general, grassland forms a transitional belt between areas of desert and forest. They are regions of intermediate and highly seasonal rainfall where there is sufficient moisture to support a drought-resistant vegetation of grasses, shrubs and in some cases trees.*

Between the fierce aridity of the desert belt and the constant humidity of the tropical forest regions lies an area where the rainfall is intermittent and erratic. The dominant plants are grasses and the habitat is one of open plains with scattered scraps of brush and woodland. As the region lies wholly within the tropics, the sun is therefore directly overhead at any one place on two occasions each year. Most of the rainfall comes at these times because the tropical convergence of global winds and the wet conditions associated with them move north and south with the sun twice every year. The intervening dry seasons are due to the dry high-pressure belt associated with the deserts moving over the grassland area.

The dominance of grasses over trees has more to do with the level of soil moisture than it has to do with the total amount of rainfall. Typically only the upper soil layers contain water, whereas the lower strata, where tree roots would be found, remain dry all year. Some areas of grassland receive substantial amounts of rain, but because it occurs only at certain times of the year trees are unable to establish themselves.

Because of the general dryness of the region it is highly susceptible to fire. Indeed it is repeated destruction by frequent grassland fires that has produced the plants that are characteristic of the habitat. The trees are particularly hard and fire-resistant and the grasses grow from their bases rather than from the tips of their leaves and stalks. They also spread by means of underground runners, which allows their instant recovery after fire has swept the area and destroyed the exposed parts.

The rapid recovery of trees and grasses from damage permits the grasslands to support large numbers of grass-eating animals despite the frequent fires. Only the upper parts of the leaves and stems are eaten, leaving the growing bases and the underground runners intact. Another feature of the grasslands that has an important influence on the fauna is the sparseness of cover. A grazing animal can therefore be seen by a predator from a great distance, and conversely the grazing animal can see danger coming. Hence both grazers and predators in these regions are highly adapted for speed and pursuit, and have long legs and quick reactions. Some birds, too, have found that they can survive on the grasslands using only their legs to take them out of trouble without recourse to flight.

Another feature of tropical grassland life is migration. Because of the seasonality of rainfall, different areas of the grasslands provide food at different times of the year and hence great migrations of grazing herds occur throughout the year. Migration also takes place to the grasslands from other parts of the world. Many birds summer in the temperate woodlands, far to the north, and fly south to the grasslands to escape the winters.

As with the desert belts, the total land area of the globe lying within the tropical grassland climatic belt has diminished since the Age of Man due to the constant northward movement of the Australian continent. Although the largest extent of tropical grassland is on the African sub-continent, considerable expanses also exist on the South American island continent south of the equatorial rain forests.

The flightless guinea fowl is found in the African sub-continent south of the equator. It is one of the fiercest and most territorial of all tropical ground-dwelling birds.

FLIGHTLESS GUINEA FOWL
*Pseudostruthio gularis*

Except for the colour of their legs there is little difference between males and females. Males have pink legs and females have blue.

When threatened, the guinea fowl inflates its throat pouch, arches its neck far back until it nearly touches its back and utters a shrill, throaty scream.

Mating occurs in the early summer and normally results in one or two eggs being laid five or six weeks later. The incubation is shared by both males and females.

80

# THE GRASS-EATERS

*Ground-dwelling birds and herbivorous herds*

The grasslands, both tropical and temperate, are the home of the running animals. The long vistas and the general lack of cover make concealment difficult and speed is the most practical means of defence.

The grasslands first appeared on a large scale about 80 million years ago, when a general reduction of global temperatures, causing a drop in average rainfall, produced a reduction in the area of forest found on the earth. At this time the mammals that had been in existence for about 20 million years developed running forms in large numbers for the first time. The grasses, representing a vast untapped food source, had however a high silica content, which made them much tougher than the leaves of trees to which the browsing mammals were accustomed. To deal with this more fibrous material new tooth structures appeared that had hard enamel ridges and cusps to grind down the grass before it was swallowed. New, elaborate digestive systems were also evolved to deal effectively with it.

By the Age of Man, the long-legged grazers, the ungulates such as the zebra, *Equus*, and the gazelle, *Gazella*, were the most successful animals of the tropical grasslands. However the rabbucks, which originated in the temperate woodlands, after man's extinction, spread southwards, round the mountain barriers, into the African and Indian sub-continents, where they flourished and competed so effectively with the ungulates that in time they largely replaced them.

Although many forms of rabbuck inhabit the same region, because of their different feeding habits they do not compete directly with one another. The little picktooth, *Dolabrodon fossor*, feeds on low-growing herbs and roots, which it digs up with its tusks and spurs. Its second incisor teeth are developed into strong laterally directed tusks and it has long spur-like claws on the fourth digit of each forefoot. As it runs only on the second and third toes of each foot, the spurs do not hinder it.

The taller grasses are grazed by vast herds of larger rabbucks, *Ungulagus* spp. They tend to be similar to their temperate cousins, but are on the whole lighter in build and have longer legs and ears. Their coloration is very different, consisting mostly of pale brown and white arranged in stripes or spots depending on the species. The strank, *U. virgatus*, has a dazzling pattern of stripes like the extinct zebra, while the larger watoo, *U. cento*, carries large angular blotches similar to those once possessed by the giraffe. Such patterns make individuals merge into one another so that a distant predator gets only a confused impression of the herd as a whole. It is particularly effective in thorn thickets and areas of scrubby woodland. All rabbucks, whether temperate or tropical, retain the dazzling white tail of their rabbit ancestors. It is used as a warning signal when the herd is attacked.

The tropical grasslands are the home of a species of large flightless guinea fowl, *Pseudostruthio gularis*. Standing about 1.7 metres high, it sports a startling selection of erectile wattles and inflatable throat pouches, which are used in threat displays when dominance or pecking order is threatened. It is an omnivorous bird and feeds on seeds, grasses, insects and small reptiles. Although it can deal a lethal blow with its broad feet in common with most plains-dwelling animals it runs off when real danger threatens.

*The picktooth*

*Only two of the picktooth's toes are functional. The fourth toe on each of its front feet has developed into a spur.*

*The picktooth's skull is similar to a rabbit's. Its tusks are developed from the front incisors.*

*The strank's stripes produce a confused impression from a distance.*

*The strank*

*The watoo*

*Blotched markings give the watoo camouflage in areas of scrubby woodland*

# GIANTS OF THE PLAINS

*The place of large herbivores in a tropical environment*

The long-necked gigantelope is a browser and eats the leaves and shoots of trees.

It has two vestigial horns, no more than bony pads on the top of its head.

Unlike most members of the gigantelope family, long-necked gigantelopes are not herd animals. They are typically found in ones and twos in lightly wooded areas around the margins of the tropical forests.

The elephants flourished throughout the first half of the Age of Mammals, but with man's appearance their numbers fell until they had almost become extinct. Two genera only, *Elephas* and *Loxodonta*, were latterly contemporaries of man and both of these died out shortly before man's disappearance, leaving no descendants. The ecological niche which they vacated was eventually filled by the descendants of a surviving group of antelopes, the gigantelopes. These enormous creatures with tree-trunk legs and weighing up to ten tonnes became the giant herbivores of the tropical plains, a group of animals feeding on trees, grasses or roots depending on the species. They had long since abandoned the antelope's running gait and had instead taken up a plodding existence – the two-toed feet of their ancestors having become broad-hooved pads.

The typical grassland-dwelling type, *Megalodorcas giganteus*, has four horns – one pair curving down behind its ears and another pair pointing out in front of its snout. Each horn has a pick-like point, enabling the animal to scrape soil away from the plant roots and bulbs on which it feeds.

The animal's basic shape was highly successful and in the course of time the gigantelopes spread northwards from tropical Africa, crossing the Himalayan Uplands in two separate waves of migration; one spreading into the coniferous forests and giving rise to the hornheads, *Cornudens* spp., and the other, much later, reaching the tundra and providing the ancestors of the woolly gigantelope, *Megalodorcas borealis*.

Once the massive body of the gigantelope had been established a number of variations appeared. The earliest was the long-necked gigantelope, *Grandidorcas roeselmivi*, a gigantelope able to browse on twigs and branches 7 metres above the ground, well out of reach of the smaller herbivores and even of its own massive cousins. As well as a long neck this animal also has a long, narrow head, enabling it to push its thick muscular lips between the branches of the trees to reach the tastiest morsels. The horns of its ancestors are reduced to long, low, bony pads at the top of the skull. Anything more elaborate would become entangled in the branches.

At first glance these massive beasts seem to contradict the general rule that animals of hotter climates tend to be smaller than their equivalents in cooler areas. The larger an animal is, the smaller its surface area is in relation to its body mass, and the more difficult it is for it to lose excess heat. In the case of gigantelopes, however, this problem is overcome by the possession of a large dewlap beneath the neck, which is well served with blood vessels and effectively increases the creature's body area by about a fifth, thus providing an efficient heat radiator.

The rhinoceros, another of the massive tropical grassland animals that became extinct during the Age of Man, has an almost direct equivalent in the gigantelopes – the rundihorn, *Tetraceras africanus*. It has adopted a body size and a horn arrangement not unlike its predecessor's and is a grazing animal, a fact that is reflected by its broad snout and muzzle. Its alarming horn array is used for defence, although the animal has few enemies likely to risk a frontal attack. For the males, however, its secondary function – for sexual display – is now more important.

The gigantelopes' immediate ancestors had long double-pointed antler-like horns

In some, the rear portions disappeared, leaving long forward-pointing prongs.

Until recently the shovel-horned gigantelope was found on the grasslands. It is thought to have lived near rivers and lakes and to have fed mainly on water plants

82

The gigantelope and rundihorn are representative members of a family of tropical heavy grazing animals that are descended from the now extinct antelopes

RUNDIHORN
*Tetraceras africanus*

GIGANTELOPE
*Megalodorcas giganteus*

The gigantelopes, like the previous occupants of this ecological niche, the elephants and rhinos, are purely herbivorous.

# THE MEAT-EATERS

*Predators and scavengers of the plains*

The raboon is directly descended from the baboon. It has evolved a bipedal stance and much heavier hind quarters.

Male raboons are larger than females and only the males have manes. Their teeth follow the general carnivore pattern.

Female

Male

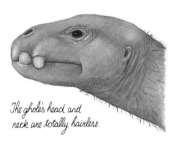

The ghole's head and neck are totally hairless.

Its massive canine teeth and molars are designed for breaking and crushing bones

Gholes frequently devour their food beneath the shelter of an overhanging termite mound, where they find protection from the sun. The termites in return feed on the remaining scraps.

Although the two principal predators of the tropical grasslands of the African sub-continent are both primates, they have evolved along very different lines and hunt different prey.

The horrane, *Phobocebus hamungulus*, is descended from the tree-dwelling apes of the tropical forests, a fact indicated by the way that the animal walks on the knuckles of its forefeet. It leads a totally ground-dwelling carnivorous mode of life. Lying in the long grass, where it is camouflaged by its stripes and mane, it waits for its chief prey, the gigantelopes. As they pass by, the horrane leaps out on to the back or neck of its quarry, using its sickle-like claws to rip deep wounds around the neck and throat. Severely wounded, the gigantelope soon dies, providing a meal for the whole horrane family group.

The other main predator is the raboon, *Carnopapio* spp. Descended from the baboons that flourished on the grasslands during the Age of Man, their diet changed from omnivorous to carnivorous during the period that the big cats, of the grasslands, died out. At the same time they increased their speed by taking to their hind limbs and adopting a totally bipedal locomotion. The forelimbs became reduced and the head was carried further forward, balanced by a thick, heavy tail. In physical form the raboon bears a distinct resemblance to the carnivorous dinosaurs that died out more than a hundred million years ago.

A number of species of raboon, each living on a different species of prey, exist in family-based tribes, like the ancestral baboons. *Carnopapio longipes* is a very small, lightly built species about 1.8 metres high that hunts smaller animals. *C. vulgaris* is the most widely ranging species and preys on the rabbuck herds. *C. grandis* is the most massive member of the genus. It stands about 2.3 metres high at the hip and lives purely as a scavenger. As predators such as the horrane eat only the softer tissues and muscles of the gigantelope's belly and anal regions there is always plenty of meat left for the scavengers. The giant raboon concentrates on the meat of the limbs and neck, leaving the rest to smaller, less powerful carrion feeders.

The most efficient scavenger of the African grasslands is the ghole, *Pallidogale nudicollum*, a creature that resembles a large mongoose. Its head and neck are almost totally devoid of hair, allowing it to reach inside the body cavities of carcases without its coat becoming fouled. Its canine teeth are particularly huge and are capable of crushing most bones to get at the marrow. Gholes live in packs of about a dozen and have developed an almost symbiotic relationship with a species of termite. This termite builds its mound with a horizontal shelf projecting out all round, a metre or so above the ground. The shelf provides shelter from the fierce midday sun where the ghole can bring bones and other tough parts of its meal to chew at leisure. The termites feed on the scraps of carrion that the ghole invariably leaves scattered around the mound, thus benefiting from the relationship. It usually takes about three days for the predators and scavengers of the grasslands to reduce a gigantelope to no more than a few pieces of bone and hide and a patch of stained, trampled ground. The final remnants are consumed by insects and micro-organisms.

# HORRANE
*Phobocebus hamungulus*

The gigantelope's principal predator is the horrane, a predatory primate. The horrane's hunting strategy relies heavily on concealment and the element of surprise.

*Carnopapio grandis*

*Carnopapio vulgaris*

# RABOONS
*Carnopapio spp.*

The raboons form a group of predatory and carnivorous animals. The largest, *Carnopapio grandis*, is a scavenger. More powerful than the horrane it feeds on the tougher parts of the carcase.

*Carnopapio longipes*

The next link in the chain of predators and scavengers is formed by the gholes. They are unable to compete with the raboons and must wait until they have finished.

# GHOLE
*Pallidogale nudicollum*

Gholes feed mainly on the kill's skeletal remains. They crack open the bones to get at the marrow, which provides them with most of their nutritional needs.

# TROPICAL FORESTS

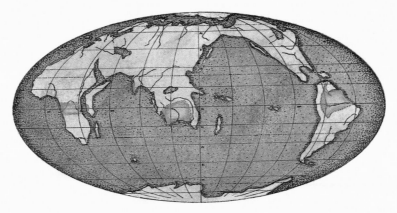

*Tropical forest is found in equatorial latitudes, where converging air currents bring large quantities of rain to the region at all seasons. This, combined with the constant high temperature, produces the forest's characteristic luxuriant vegetation.*

The tropical forests are found in a broad belt encircling the world at the equator, broken only by oceans and mountains. Their distribution coincides with the band of low-pressure areas that occurs where rising tropical air is replaced by moist air flowing in from the north and south to form a system of converging winds.

The rain forest is the floral product of great heat and copious moisture. At all times the average temperature must be between about 21°C and 32°C and the annual rainfall in excess of 150 centimetres. As the sun is roughly overhead throughout the year, the climatic conditions have a constancy found in no other habitat.

Tropical forests are often associated with great rivers, which carry away the copious rainfall. Such rivers are found in the South American island continent, the African sub-continent and the sub-continent of Australia.

Despite the constant fall of discarded leaves the soils of the rain forests are very thin. The conditions are so favourable for decomposition that humus does not have a chance to form. The tropical rain washes the clay minerals out of the soil, preventing important nutrients such as nitrates, phosphates, potassium, sodium and calcium from being retained as they are in temperate soils. The only nutrients found in tropical soils are contained in the decomposing plants themselves.

There are many variations on the basic form of tropical forest resulting both from climatic and local environmental differences. Gallery forest is found where the forest comes to an abrupt halt, as at the edge of a broad river. Here the branches and leaves form a dense wall of vegetation reaching to the ground, to take advantage of light coming in from the side. Less luxuriant monsoon forests exist in regions where there is a distinct dry season. They are found at the edge of continental areas, where the prevailing winds blow from the dry interior at one particular time of year, and are typical of the Indian peninsula and parts of the Australian sub-continent. Mangrove forest is found in saline swamp areas along muddy shorelines and the mouths of rivers.

There are no dominant species of trees in the tropical forest as there are in other forest habitats. This is because there are no seasons and therefore the insect population does not fluctuate; the insects that feed on a particular species of tree are always present and will destroy the seeds and seedlings of that tree if they are sown nearby. Therefore the only seeds that flourish are those that are transported some distance away from their parent and its permanent insect population. In this way stands of particular tree species are prevented from forming.

The area of tropical forest has increased considerably since the Age of Man. In the past a great deal of damage was done to the habitat by man's agricultural practices. Primitive societies cut down areas of trees and farmed the clearings for a few years until the thin soil became exhausted, compelling them to move on to another area. In the cleared areas the original forest did not immediately re-establish itself and it was many thousands of years after man's extinction before the tropical forest belt returned to anything like its natural condition.

# THE TREE-TOP CANOPY

*A world of gliders, climbers and perchers*

The long-armed ziddah.

The ziddah curls itself into a ball to sleep. First it wraps its arms across its body and then brings its legs close in against its chest.

The tropical forest is one of the most luxuriant habitats on earth. The high rainfall and stable climate mean that there is a perpetual growing season and there are therefore no periods in which there is nothing to eat. The copious vegetation, thrusting upwards to reach the light, although continuous, is arranged very roughly in horizontal layers. Most photosynthesis takes place at the very top, in the canopy layer, where the tops of the trees branch out to form an almost continuous blanket of greenery and flowers. Beneath this the sunlight is more diffused and the habitat consists of the trunks of the taller trees and the crowns of those that do not quite reach the canopy. The forest floor is the gloomy domain of shrubs and herbs, which sprawl out to make the best use of the little light that filters down.

Although the tremendous variety of plant species supports an equal diversity of animal species, the number of individuals in each is comparatively small. This situation is exactly the opposite of that found in harsh environments such as the tundra, where, because few life forms can adapt to the conditions of the region, there are many fewer species of either plants or animals but correspondingly more individuals in each. As a result the animal population of the tropical forests remains stable and there are no cyclical plagues of either predator or prey species.

Birds of prey such as eagles and hawks are the important predators of the tree tops, as they are in any other habitat. The tree-living animals of these regions must be swift enough to elude them and also to escape from tree-climbing predators coming up from below. The mammals that accomplish this best are the primates – the monkeys, apes and lemurs. The long-armed ziddah, *Araneapithecus manucaudata*, of the African sub-continent has taken these specializations to the extreme, and has developed long arms and legs, fingers and toes, so that it can brachiate, or swing, its tiny globular body through the branches of the trees at high speed. It has also evolved a prehensile tail, just as its South American cousins did in the first half of the Age of Mammals. Its tail, however, is not used for locomotion but only for hanging from when resting or asleep.

The flunkey, *Alesimia lapsus*, a very small marmoset-like monkey, has become adapted to a gliding mode of locomotion. In this development it parallels the evolution of many other mammals that have evolved gliding wings, or patagia, from folds of skin between the limbs and tail. To support the patagia and deal with the stresses involved in flight the backbone and the limb bones have become remarkably strong for an animal of this size. Steered by its rudder-like tail the flunkey makes great gliding leaps between the crowns of the highest trees to feed on fruit and termites.

Among the tree-living reptiles of the African rain forest perhaps the most specialized is the anchorwhip, *Flagellanguis viridis* – an extremely long and thin tree snake. Its broad, grasping tail, the most muscular part of its body, is used to anchor it to a tree while it lies coiled and camouflaged among the leaves of the tallest crowns in wait for an unwary passing bird. The snake is capable of darting out three metres, equivalent to about four-fifths of its body length, and seizing its prey while still retaining a tight hold on the branch with its tail.

Coiled against a tree, the anchorwhip lies in wait.

With jaws gaping it shoots out to seize a passing bird in mid-flight.

88

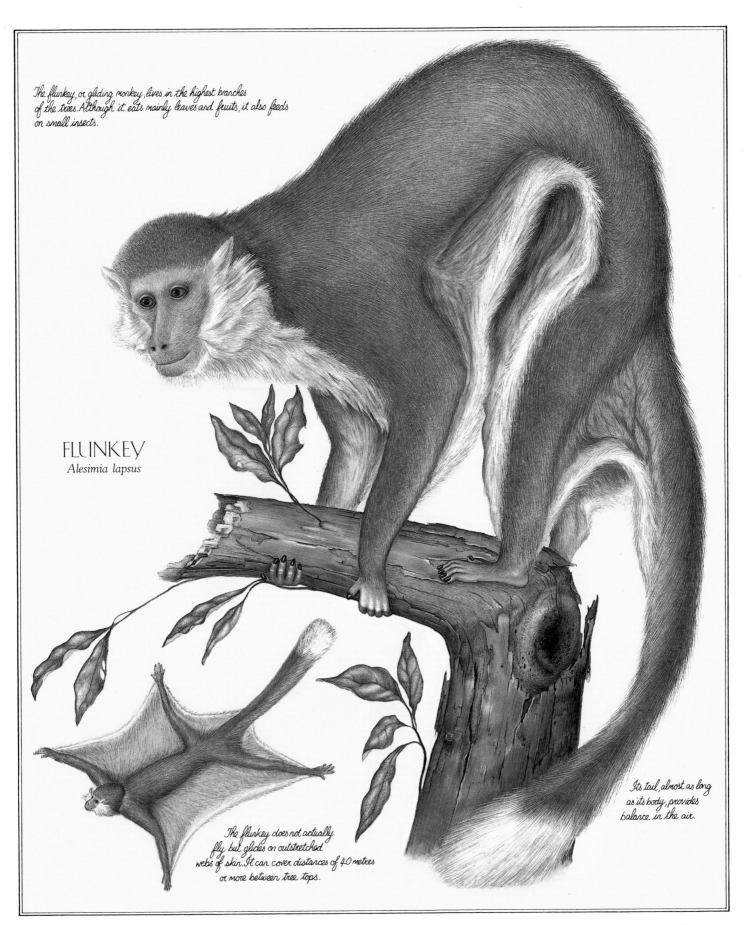

The flunkey, or gliding monkey, lives in the highest branches of the trees. Although it eats mainly leaves and fruits, it also feeds on small insects.

# FLUNKEY
*Alesimia lapsus*

Its tail, almost as long as its body, provides balance in the air.

The flunkey does not actually fly but glides on outstretched webs of skin. It can cover distances of 40 metres or more between tree tops.

89

Male khiffahs are equipped primarily for defence.

Unlike most members of the cat family, the striger has grasping claws.

KHIFFAH
*Armasenex aedificator*

The striger is the khiffah's most important predator. Climbing like a monkey it can easily reach the khiffah's tree-top lair.

STRIGER
*Saevitia feliforme*

A pad of hairless skin at the tip of the tail is used for gripping branches.

# LIVING IN THE TREES

*The evolution of life under threat*

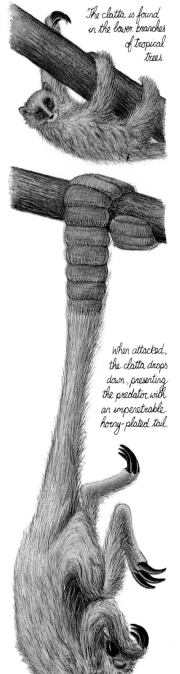

The clatta is found in the lower branches of tropical trees.

When attacked, the clatta drops down, presenting the predator with an impenetrable horny-plated tail.

During most of the Age of Mammals the apes and monkeys enjoyed a degree of security among the tree tops. For even though there were some predators, none was adapted to prey on them specifically – but that was before the striger.

This fierce little creature, *Saevitia feliforme*, developed from the last of the true cats about 30 million years ago and spread throughout the rain forests of Africa and Asia, its success hinging on the fact that it was as well adapted to life in the trees as its prey. The striger even adopted the bodily shape of the monkeys on which it fed; a long, slender body, forelimbs that could swing apart to an angle of 180°, a prehensile tail and opposable fingers and toes that allowed it to grasp the branches.

With the coming of the striger the arboreal mammal fauna of the tropical forest underwent considerable change. Some of the slow-moving leaf- and fruit-eating animals were wiped out completely. Others, however, were able to adapt in the face of this new menace. As usual, when an environmental factor as radical as this is introduced, evolution takes place in a rapid leap, because now quite different physical attributes are advantageous.

The clatta, *Testudicaudatus tardus*, a lemur-like prosimian with a heavily armoured tail protected by a series of overlapping horny plates, demonstrates this principle. Before the arrival of the tree-living predators, such a tail would have been a disadvantage, interfering with the efficiency of food gathering. Any tendency for such a cumbersome structure to evolve would have been quashed rapidly by natural selection. But faced with constant danger the efficiency of food gathering would have taken on an importance secondary to defence and would have therefore created the correct conditions for it to evolve.

The animal itself is a leaf-eater and moves slowly, upside down, along the boughs. When a striger attacks, it drops down and hangs from a branch by its tail. The clatta is now safe – the only part within reach of the predator is too heavily armoured to be vulnerable.

The khiffah, *Armasenex aedificator*, is a monkey whose defence is based on its social organization. It lives in tribes of up to twenty individuals and builds defensive citadels in the boughs of trees. These large, hollow nests, woven from branches and creepers and roofed with a rainproof thatch of leaves, have several entrances, usually situated where the main branches of the tree thrust through the structure. Most of the work of food gathering and building is carried out by females and young males. The adult males remain behind to defend the citadel and have developed a unique set of features to carry out their highly specialized role; horny armour over the face and chest and vicious claws on the thumb and forefinger.

It is not unknown for a female to taunt a passing striger and allow herself to be pursued back to the citadel, dashing to safety while the striger finds its way barred by a powerful male capable of disembowelling it with a swipe from its terrible claws. This apparently senseless behaviour, however, provides the colony with fresh meat, a welcome supplement to their basic vegetarian diet of roots and berries. Only young and inexperienced strigers are caught this way.

Female and young male khiffahs possess neither armour nor claws. They are the colony's principal food gatherers.

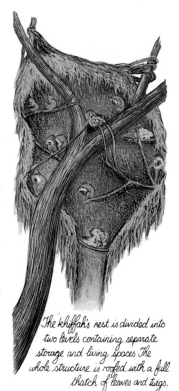

The khiffah's nest is divided into two levels containing separate storage and living spaces. The whole structure is roofed with a full thatch of leaves and twigs.

# THE FOREST FLOOR

*The twilight zone of woodland life*

Clinging to the trunks of trees with their clawed fingers, trovamps are ideally placed to leap, dart-like, on-to their prey.

On each jaw the trovamp has two barbs, formed from the canine teeth. When the jaw is closed they protrude to give the appearance of tusks.

Compared with the canopy layer the floor of the tropical rain forest is a dark, humid place. Little light penetrates through from the tree tops, and although there are many shrubs and herbs they nowhere present a thick, impenetrable barrier. Despite a steady fall of dead leaves from above, the soil cover is very shallow. The vegetable material on the ground is under constant attack from micro-organisms and from the ubiquitous termites that perform a broadly similar function to that of the earthworms in temperate latitudes by keeping the debris circulating.

These termites are the principal food source of the turmi, *Formicederus paladens*, one of the few large mammals found on the African rain forest floor. It is descended from the pigs that were once common in this environment. In the turmi the tusks of the upper jaw are projected forwards, elongating the snout still further, and have turned outwards to produce strong pick-like instruments with which it digs into termite mounds. The lower jaw has lost all its teeth and musculature and the mouth has diminished to a tiny hole through which sticks out its ribbon-like tongue to gather termites.

From the same ancestral pig stock as the turmi comes the zarander, *Procerosus elephanasus*. This much larger vegetarian animal lives on the sparse herbs and shrubs found in less dense areas of the forest floor. Its long trunk, developed from a snout similar to the trunk of the ancient elephant, enables the zarander to reach leafy branches 4 metres above the ground, where it can snip branches and vines from the trees by the scissor action of its upper and lower tusks. Despite its long nose, the zarander has little sense of smell. Like other mammals of the forest floor, the lack of wind and general circulation among the dense trees means that scents do not travel far. Relying on its keen hearing to warn it of the approach of an enemy, it takes off into the thicker parts of the forest at the arrival of a predator, squeezing its narrow body between the tree trunks, and remaining motionless, camouflaged by its stripes and dark body colour.

One of the smaller mammals of the African tropical forest is the trovamp, *Hirudatherium saltans*, a parasite which sucks the blood of larger mammals. The trovamp is built rather like an insectivore or one of the smaller prosimians. It is very agile, climbing about, usually in packs, among the trunks and the branches of the shrubs. The trovamp is a prodigious jumper and can leap 3 metres from a branch to bury its needle-like jaws into the hide of a passing animal. Its protruding canine teeth act as barbs and prevent it from being dislodged from its host until it is finished feeding. As many as ten trovamps may parasitize one host and will remain feeding until the animal is severely weakened.

A large number of birds inhabit these regions. The most remarkable from the point of view of social behaviour is the giant pitta, *Gallopitta polygyna*. The male pitta, unusually for a bird, is about three times the size of the female and each year takes a harem of three or four females. Each female builds a separate nest in the same vicinity and relies on the male to provide her with food during the breeding season. The male also provides protection from predators as well as defending the harem against rivals.

A single animal may be parasitized by as many as ten trovamps at one time. Each trovamp is held fast to the creature's side by barbed teeth and curved front claws.

Male pittas guard their harem of females throughout the breeding season. Each female occupies a separate nest.

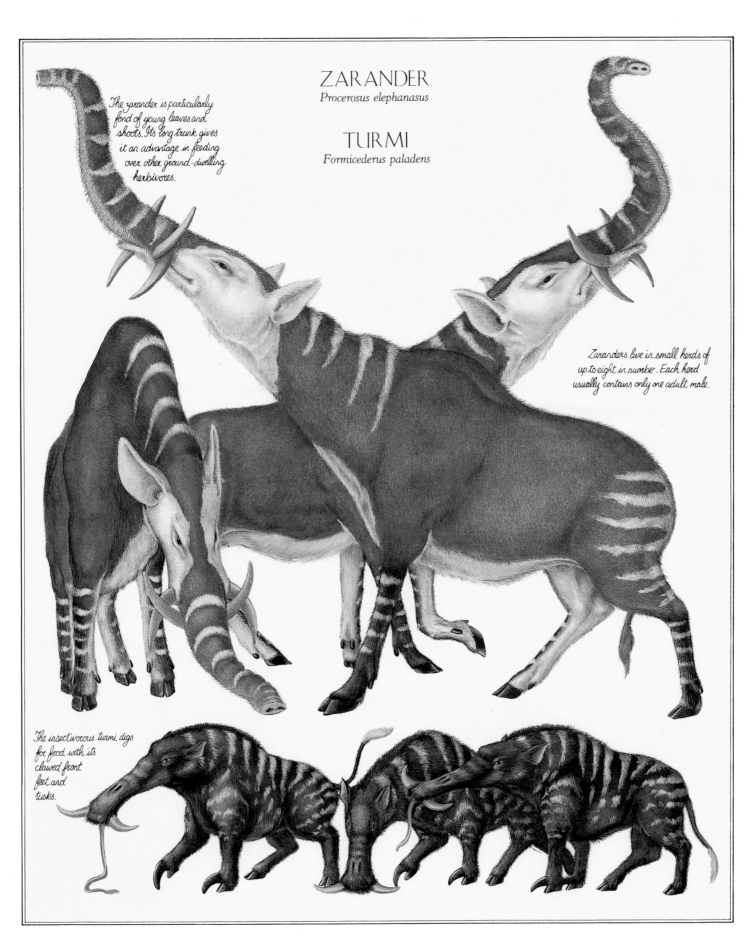

# ZARANDER
*Procerosus elephanasus*

# TURMI
*Formicederus paladens*

The zarander is particularly fond of young leaves and shoots. Its long trunk gives it an advantage in feeding over other ground-dwelling herbivores.

Zaranders live in small herds of up to eight in number. Each herd usually contains only one adult male.

The insectivorous turmi digs for food with its clawed front feet and tusks.

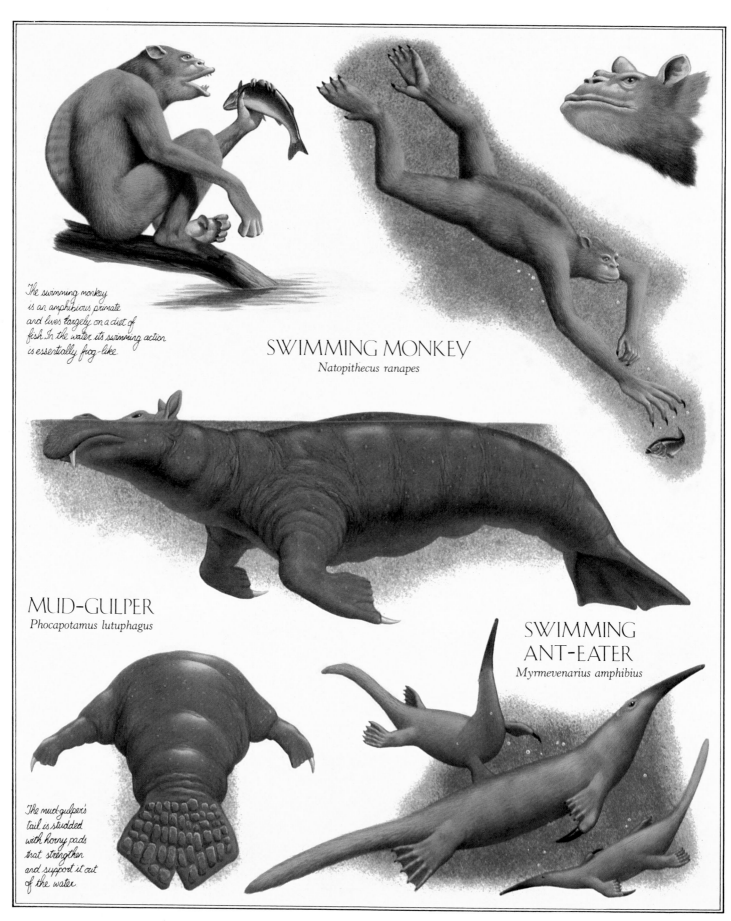

The swimming monkey is an amphibious primate and lives largely on a diet of fish. In the water its swimming action is essentially frog-like.

## SWIMMING MONKEY
*Natopithecus ranapes*

## MUD-GULPER
*Phocapotamus lutuphagus*

## SWIMMING ANT-EATER
*Myrmevenarius amphibius*

The mud-gulper's tail is studded with horny pads that strengthen and support it out of the water.

94

# LIVING WITH WATER

*Creatures of the tropical wetlands*

The mud-gulper lives largely on water plants which it dredges from the muddy bottoms of rivers and lakes. On land the mud-gulper tucks its tail under its body.

As a signal to the opposite sex, the toothed kingfisher's beak changes colour early in the breeding season.

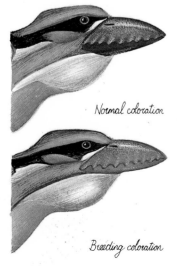

*Normal coloration*

*Breeding coloration*

The largest aquatic mammal of the African swamplands is the mud-gulper, *Phocapotamus lutuphagus*. Although it is derived from a water-dwelling rodent it shows adaptations that closely parallel those of the extinct ungulate hippopotamus. Its head is broad and its eyes, ears and nostrils are located on bumps on the top so that they can still operate even when the animal is totally submerged. The mud-gulper eats only water plants, which it scoops up in its wide mouth or scrapes up from the mud with its tusks. Its body is long and its hind feet are fused to form a flipper, giving it a seal-like appearance. Even though it is very clumsy out of the water it spends much of its time on mudbanks, where it breeds and rears its young in noisy colonies at the water's edge.

Less well adapted but nevertheless efficient in the water is the swimming monkey, *Natopithecus ranapes*. Descended from the swamp monkey, *Allenopithecus nigroviridis*, of the Age of Man, this creature has developed a frog-like body with webbed hind feet, long, clawed fingers for catching fish and a ridge down its back to give it stability in the water. Like the mud-gulper, its sensory organs are placed high up on its head. It lives in riverside trees, from which it dives to catch the fish that are its staple diet.

Land-dwelling animals that have taken to an aquatic mode of life have usually done so initially to escape land-dwelling predators. This is probably why the water ant has taken to building its huge nest on rafts in swamps and quiet backwaters. Each nest is made of twigs and fibrous vegetable material, waterproofed by a plaster of mud and bodily excretions. It is connected to the banks and to floating foodstores by a network of bridges and ramps. However, in their new mode of life the ants are still vulnerable to the swimming ant-eater, *Myrmevenarius amphibius*, which has evolved in parallel with it. The ant-eater lives solely on the water ants, and to reach them undetected it attacks the nest from below, ripping through the waterproof shell with its clawed paddles. Since below the waterline the nest is made of discrete chambers that can rapidly be made watertight in an emergency, little damage is done to the colony as a whole. The ants drowned in an attack, however, are enough to feed the ant-eater.

Fish-eating birds, like the toothed kingfisher, *Halcyonova aquatica*, are frequently found along the water courses of the tropical swamps. The bill of the kingfisher is strongly serrated with tooth-like points that help it to spear fish. Although it cannot fly as well as its ancestors, nor can it hover or dive as they did, it has become adept at "underwater flight", pursuing its prey in their own medium. After catching a fish, the kingfisher brings it to the surface and gulps it into its throat pouch before taking it back to the nest.

The tree duck, *Dendrocygna volubaris*, is a water-living creature that almost seems to have changed its mind about its preferred habitat and appears to be in the process of undergoing a change back to the more arboreal lifestyle of its remote ancestors. Although it is still duck-like, the webs on its feet are now degenerate and its rounded beak more suitable for feeding on insects, lizards and fruit than on water organisms. The tree duck still takes to the water to escape predators and its young do not venture on to land until they are nearly adult.

Although of recent aquatic origin the tree duck lives mainly on land.

The toothed kingfisher is not a swimming bird in the usual sense. It uses its wings rather than its feet – a method that is particularly successful under water.

**SLOBBER**
*Reteostium cortepellium*

Separate races of marsupial slobber are found on different species of tree and creeper, each race's hair varying in texture to blend in with the surroundings. The slobber's young are carried protruding motionless from a pocket in their mother's abdomen.

**HIRI-HIRI**
*Carnophilius ophicaudatus*

# AUSTRALIAN FORESTS

*Marsupial climbers and marsupial predators*

*The chuckaboo, a marsupial monkey, is a communal tree-dweller.*

*It has a prehensile tail with a hairless gripping pad at the tip.*

*The female has two pouches, one on either side of her abdomen, so that her young do not get in the way when she is climbing.*

Beyond the mountains of the Far East – the most extensive and the highest chain in the world, greater even than the Himalayas at their zenith 50 million years ago – lies the great Australian sub-continent.

The conditions in this area today – lush tropical forests occupying vast river basins – make it difficult to believe that a mere 100 million years ago this landmass was part of the Antarctic continent. When at this time Australia split off and began drifting northwards, the Age of Mammals was well under way and the continent already had its own mammal population. These mammals were nearly all marsupials – mammals that nursed their young in a pouch on their abdomen – and because of Australia's long history of isolation have largely remained so. In the rest of the world, however, the marsupials were gradually superseded by the placentals – mammals not giving birth until their young are more fully developed.

By the Age of Man, Australia had reached the desert and tropical grassland latitudes, where the conditions provided the evolutionary impetus for the development of running and burrowing animals such as the kangaroo, *Macropus* spp., and the wombat, Vombatidae. After man the continent continued its drift northwards until, sometime in the last ten million years, it collided with the mainland, throwing up the great barrier mountains that exist today. Although some diffusion of animals has taken place between Australia and the rest of the Northern Continent, the mountains have kept this cross-traffic to a minimum and the sub-continent still has a predominantly marsupial fauna – albeit one adapted to the tropical forests.

As in previous ages the Australian marsupials have developed forms that are superficially very similar to those of placental creatures existing in similar environments in other parts of the world. A prime example of this is the chuckaboo, *Thylapithecus rufus* – essentially a marsupial monkey with grasping arms and legs, opposable digits and a prehensile tail. Its bodily form, similar to many of the true monkeys in other parts of the world, is well suited to life in the trees.

A less energetic tree-dweller, the slobber, *Reteostium cortepellium*, can be thought of as a kind of marsupial sloth that spends nearly all of its life hanging upside down from trees and creepers. It is totally blind and subsists entirely on insects that it catches in the flowers of its home creeper by entangling them in long strands of mucus dangled from its mouth. Its large downturned ears and sensory whiskers alert it to an insect's arrival and tell it when to drop the mucus, which it aims at the flower's scent. As the slobber's hair grows in spiral tufts and is pervaded by a parasitic algae, it is completely camouflaged against the background of creepers, and when totally motionless can escape the attention of predators.

A marsupial predator that the slobber takes pains to avoid is the hiri-hiri, *Carnophilius ophicaudatus*, which, despite the fact that it is a tree-dweller, is also highly efficient in preying on ground-living animals. Lying in wait on a low branch, it dangles its strong prehensile tail down like an innocent vine. When some unsuspecting animal trots by, the hiri-hiri seizes it swiftly with its tail and strangles it. The hiri-hiri is descended from the Tasmanian devil, *Sarcophilus harrisii*.

# THE AUSTRALIAN FOREST UNDERGROWTH

*Life on the forest floor*

The floor of the great rain forest of the Australian sub-continent is home for a number of marsupial mammals. One of the most generalized and successful of these is the omnivorous posset, *Thylasus virgatus*, the marsupial equivalent of the tapir. Like its placental counterpart, it wanders through the gloomy undergrowth in small herds, snuffling and scraping for food in the thin soil with its flexible, sensitive snout and protruding tusks. Cryptic coloration helps to conceal it from its enemies.

The largest animal of the Australian forest, and in fact the largest animal found in any of the world's tropical forests, is the giantala, *Silfrangerus giganteus*. This animal has evolved from the plains-dwelling kangaroos and wallabies that were common when much of the continent was semi-arid grassland, and betrays its ancestry by its upright stance and peculiar loping motion. The giantala is so large that it seems at first sight ill-adapted to life in the confined conditions of the tropical forest floor. However, its great height does give it the advantage that it can feed on leaves and shoots that are well out of reach of the other forest inhabitants and its bulk means that shrubs and small trees do not impede it. As the giantala crashes through the thickets, it leaves behind well-marked trails, which, until they are reclaimed by the natural growth of forest, are used as trackways by smaller animals such as the posset.

Convergent evolution on the Australian sub-continent is not solely characteristic of the marsupials. The fatsnake, *Pingophis viperaforme*, descended from one of the many elapid snakes that have always been a feature of Australian fauna, has adopted many of the characteristics of forest ground-dwelling viper snakes such as the gaboon viper and puff adder of the long-lived genus *Bitis* that are found in other parts of the Northern Continent. These include a fat, slow-moving body and a coloration that renders it totally invisible in the leaf litter of the forest floor. The fatsnake's neck is very long and slender and allows its head almost to forage independently of its body. Its main method of catching prey is to deal it a poisonous bite from where it lies hidden. Only later, when its venom has finally killed it and begun its digestive function, does the fatsnake finally catch up with it and eat it.

Australian bower birds have always been noted for the fantastic structures built by the male for the purpose of wooing a female. The hawkbower, *Dimorphoptilornis iniquitus*, is no exception. The bower itself is quite a modest affair, housing the permanent nest and a small altar-like structure at the entrance. While the female incubates the eggs, the male, a rather hawk-like bird, catches a small mammal or reptile and sets it on the altar. The offering is never eaten but serves as bait to attract flies, which are then caught by the female and fed to the male to ensure his continuing attention during the long incubation period. Once the eggs have hatched the chicks are fed on the fly larvae that have developed in the rotting carrion.

Another curious bird is the termite burrower, *Neopardalotus subterrestris*. This mole-like bird lives entirely underground in termite nests, where it digs nesting chambers with its huge feet and feeds on the termites with its long and sticky tongue.

*The poisonous fatsnake can strike out at prey 5 to 10 metres away from where it is lying.*

*The fatsnake's body is heavy and slug-like.*

*The male hawkbower is more lightly built than the female.*

*The male skewers its prey outside the bower to attract flies.*

*The termite burrower is a wingless bird. Its feathers are fine and hair-like, and its long claws and shovel-shaped beak are designed for digging into termite mounds.*

*Its tongue has a bristle-like tip.*

The herbivorous giantala plucks at the leaves of trees with its long tongue. When drawn erect a fully mature individual stands over 3 metres high.

The posset forages in the soil with its long snout for roots and grubs.

# GIANTALA
*Silfrangerus giganteus*

The giantala's muscular tail acts as an extra leg, bearing the creature's weight when standing upright.

# POSSET
*Thylasus virgatus*

The posset, a forest-dwelling marsupial pig, has four tusks, two projecting from its upper jaw and two from its lower jaw.

# ISLANDS AND ISLAND CONTINENTS

*The most important isolated environments on earth lie on the South American continent and on the oceanic islands of Lemuria, Batavia and Pacaus. The accident of geographical separation has given these areas quite distinct animal communities.*

Isolation is one of the most important mechanisms of evolution. When a group of creatures becomes separated from the main breeding population, the separated group evolves independently of the parent group because there is no longer any possibility of interbreeding. The new group interacts with its environment, changing into new forms and evolving along lines that would be totally closed to it if it were living among its original enemies and competitors. This phenomenon is particularly marked where animals become isolated in sparsely or hitherto unpopulated areas and is nowhere better seen than on the islands of the oceans.

There are two main varieties of isolation in this context, each producing its own environmental pressures and giving rise to its own forms of evolution.

The first takes place when one continental mass splits away from another. What then happens to their fauna is largely dependent on the subsequent movements of the two continents. One land mass may drift northwards or southwards with respect to the other, subjecting its fauna to new climatic and environmental conditions which could affect their evolution and ultimately lead to the production of totally new genera and species. Exactly this happened during the Age of Reptiles, when the South American continent, which had shared the same dinosaur fauna as Africa, split away, resulting in the evolution of totally different animals in each area.

When a drifting continent collides with another, very often a considerable interchange of faunas takes place between the two land areas. It may happen that the fauna from one continent completely replaces that of the other. This happened when the small continent that is now the Indian Peninsula collided with mainland Asia.

The second form of biological isolation occurs when a completely new group of volcanic islands is formed. In plate tectonics much of the activity between adjacent crustal plates takes place in the open ocean. New plates are created along the mid-oceanic ridges and are destroyed as they slide beneath one another in the deep oceanic trenches. Such violent activity produces earthquakes and volcanic eruptions, creating new islands from the ocean bed.

The volcanic islands, quite barren to begin with, are soon colonized by living organisms. Plants, germinated from windblown seeds, are usually the first to arrive and take hold followed by the insects. The first vertebrate inhabitants are usually flying creatures such as birds or bats. Only later do the other vertebrates, usually reptiles and small mammals, arrive, sometimes on floating branches and tree trunks – the result of some river flood hundreds of kilometres away. All these creatures then evolve independently of their ancestral continental stock to fill all the ecological niches of the island. The classic example of this sequence is the colonization of the Galapagos Islands off the west coast of South America during the early part of the Age of Mammals. These islands were initially populated by a small number of species, which gave rise eventually to a vast array of new creatures, including four-eyed fish, marine lizards and giant tortoises. The island's fauna, particularly the inter-island differences between related species, was thoroughly investigated and stimulated the development of evolutionary theory.

# SOUTH AMERICAN FORESTS

*The effect of continental drift on animal communities*

The nightglider's erectile spines are formed from modified hairs that, in the course of evolution, have developed into stiff needle-like structures. The nightglider parachutes silently down on to its quarry, impaling it on its chest spines.

Although during the first half of the Age of Mammals South America did have a small population of primitive placental mammals, it was, like Australia, a bastion of the marsupials. However, just before the Age of Man, a land bridge was established between South America and North America which led to an exchange of faunas between the two areas. The result was that the placental mammals from the north, being more versatile, almost entirely replaced the marsupials and the primitive placentals of the south. The northern fauna were more versatile because they had been subjected to greater selective pressures in the preceding 50 million years; they had been compelled to adapt radically in the face of changing environmental conditions brought about by such factors as ice ages and faunal exchanges with Asia. The result at the time of the collision with South America was a very hardy and adaptable stock of animals. The mammals of South America, on the other hand, had experienced a stable unchanging environment during the same period and therefore lacked this essential adaptability. A similar fate did not befall the marsupials of Australia, since that continent, in drifting northwards, presented its fauna with constantly changing conditions, resulting in a population of hardy species that were able to survive the faunal exchanges that occurred during the period shortly after Australia impacted with Asia.

Twenty million years after the Age of Man the land connection with North America was again broken and South America became an island continent once more. Since the split, climatic conditions on the South American continent have remained unchanged and the fauna has therefore changed very little. This conservatism is well seen among the mammalian predators – a niche that has continued to be occupied by members of the order Carnivora despite the fact that this group has declined elsewhere.

The foremost predator of the South American tropical forest is the gurrath, *Oncherpestes fodrhami*, a giant hunting mongoose. Its ancestor, *Herpestes*, was introduced by man to the then offshore islands at the north of the continent, where it became a pest and overran them. When the islands became fused to the mainland the mongoose spread southwards and developed into its present jaguar-like form. Its chief prey is the tapimus, *Tapimus maximus*, a long-tusked rodent that feeds in open areas of the forest.

A much smaller carnivore, the nightglider, *Hastatus volans*, is derived from tree-dwelling mustellid stock. During the day it hangs on trees disguised against the bark, floating down to feed upon nocturnal insects, frogs and small mammals at night. Its method of hunting is to empale its prey on the spines that project from its chest. One of the strangest birds of these regions is the matriarch tinamou, *Gynomorpha parasitica*. The female of the species is a ground-living bird, very much larger than the male whom she carries around on her back. The male's wings and digestive system are degenerate and he is entirely parasitic on the female, sucking her blood through his needle-shaped beak. The male's only biological function is to provide sperm during mating. This relationship arises from the species' low population density, which makes it an advantage for each female to have a mate constantly available rather than to search for one each breeding season.

There are several different species of nightglider.

Each is camouflaged against particular species of forest trees.

The tiny male matriarch tinamou spends its entire life as a parasite on the back of a female.

The male has large claws on its feet and a single claw on each wing.

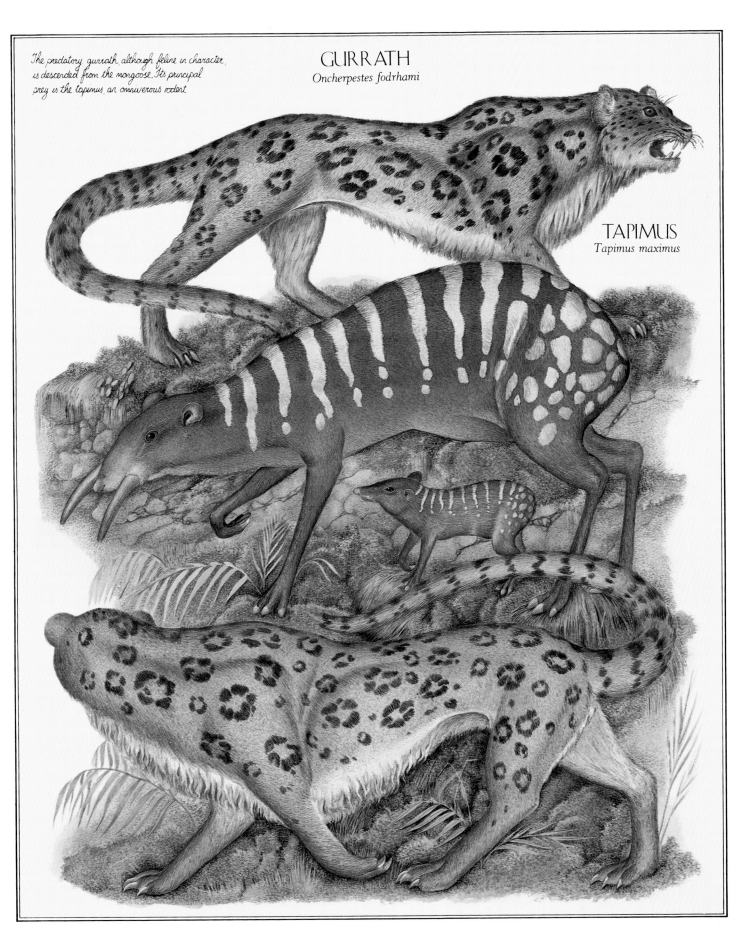

The predatory gurrath, although feline in character, is descended from the mongoose. Its principal prey is the tapimus, an omnivorous rodent

# GURRATH
*Oncherpestes fodrhami*

# TAPIMUS
*Tapimus maximus*

## STRICK
*Cursomys longipes*

Stricks are very shy creatures, and for security live in small herds of up to 12 in number. Their markings break up their shape and make them difficult to pick out at a distance.

Because stricks and wakkas occupy similar ecological niches, they have, in the course of evolution, developed along similar lines from completely different ancestors.

Wakkas are such fast runners that, provided they can see a predator approaching, they are always able to outrun it. They are only vulnerable when they have their heads down foraging in the grass.

## WAKKA
*Anabracchium struthioforme*

# SOUTH AMERICAN GRASSLANDS

*Evolution on an island continent*

The flower-faced potoo sits on the pampas with its mouth open during the middle part of the day when insects are flying.

Throughout its history, the movement of the crustal plate carrying the South American continent has been predominantly westwards, and hence the landmass has tended to remain within the same latitudes. This accounts for the constancy of the climatic regions and the conservatism of its fauna.

During the continent's early history the grasslands, or pampas, supported their own fauna of running ungulate animals, similar to, but totally isolated from, those in other parts of the world. These animals existed until the continent became joined by a land bridge to North America, when they and the native marsupial population were swept away completely by the influx of animals from the north. Strangely enough the northern ungulates did not find a permanent foothold on the pampas, but rather rodents such as the maras, *Dolichotis*, and the capybaras, *Hydrochoerus*, present at the time of man, were the more successful. In this respect the South American continent anticipated the rise of the advanced running rodents and lagomorphs in the rest of the world.

Once the continent separated from the supercontinent of the north the rodent fauna developed along its own lines. The running animals of the pampas are dominated by strange bipedal grazers, which are descended from the jumping rodents that evolved in the rain-shadow deserts along the western mountains. Although long hind legs evolved independently among desert rodents in other continents only those of South America changed from a jumping to a running mode in the course of their evolutionary history. Along with this change of gait went an increase in size and a change of dentition that effected the final transition from the jumping, gnawing rodent of the desert to the striding grazer of the plains.

The most generalized running rodent is the strick, *Cursomys longipes*, which looks very much like the grazing marsupial kangaroos that once existed in Australia. They graze among the long grasses in tightly knit groups that are large enough to ensure that there are always at least two or three individuals with their heads up, looking around for danger while the rest have their senses buried in the grass.

The most specialized creature in this family of animals, and perhaps the most highly adapted running animal in the world, is the wakka, *Anabracchium struthioforme*. Because of its bipedal stance its forelimbs have become less important and are now completely atrophied. Its globular body and long hind legs support an equally long neck and tail which balance one another, maintaining the animal's centre of gravity over its hips. These features give the creature a clear view of the surrounding countryside. Even when the wakka is grazing in long grass its eyes are placed high enough on its elongated head for it to see the approach of a predator.

The flower-faced potoo, *Gryseonycta rostriflora*, is the oddest bird found on the grasslands. The interior of its beak is coloured and patterned like the petals of a flower, so that when it has its mouth open it looks exactly like an open bloom. This elaborate mimicry is designed to deceive insects and provides the potoo with a meal by merely opening its mouth. Because tropical grassland flowers appear only when there is adequate moisture, the potoo migrates seasonally with the rains.

The strick has a small head with long ears and wide nostrils.

The strick is bipedal, running on the tips of its two-toed feet.

The wakka is also bipedal, but unlike the strick it has no front paws to help it keep its balance and instead relies on its tail.

105

# THE ISLAND OF LEMURIA

*The bastion of the hoofed animals*

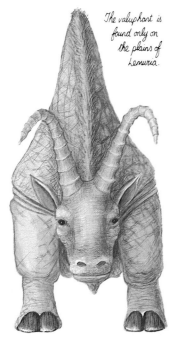

The movements of the earth's crustal plates that carry the continents and account for continental drift are the result of convection currents deep in the earth's mantle. The currents can build up stresses beneath the continents which eventually tear them apart.

Normally an elongated rift valley, associated with considerable volcanic activity, forms first of all along the line of the eventual split. The land on either side then separates and moves apart, an ocean area growing steadily to fill the gap. This happened when the small island continent of Malagasy split away from mainland Africa 100 million years ago, and again more recently when the whole of eastern Africa split away to form Lemuria.

In the case of Lemuria the separation occurred before the ungulate herds of Africa had been replaced by the rabbucks from temperate latitudes. As a result hooved animals are as plentiful on the grassy plains of Lemuria as they ever were in Africa before the Age of Man.

The valuphant, *Valudorsum gravum*, is the largest ungulate. It is a massive animal some 5 metres long with a squat, rounded body and massive legs, resembling those of the gigantelope to which it is distantly related. Its most distinctive feature is the tall ridge running down its back and neck. The ridge is supported by the neural spines of the vertebrae and may be of use in regulating the animal's temperature.

Massive horns, growing to nearly a metre in length, are the valuphant's main means of defence. Its eyes and ears are small to keep out the dust.

The valuphant feeds only on herbs and roots, which it gouges up with its horns. The grasses themselves are eaten by more lightly built fleet-footed ungulates such as the snorke, *Lepidonasus lemurienses*. The snorke has a very long head with its eyes placed near the top – an adaptation that enables it to keep a watch for predators while grazing. The upper layers of vegetation are exploited by the long-necked yippa, *Altocephalus saddi*, which can reach the leaves and young shoots of the savanna trees.

The cleft-back antelope, *Castratragus grandiceros*, a creature that is superficially similar to the ancestral antelope, has a curious symbiotic relationship with the tick bird, *Invigilator commensalis*. This relationship is really no more than a strengthening of the symbiosis that had developed between birds and grazing animals during the early part of the Age of Mammals. Birds on the grassy plains often accompanied the larger mammals, catching the insects disturbed by their hooves, or pecking ticks and mites from the hides of the animals themselves. The grazing animals tolerated this as the birds rid them of parasites and also gave warning of approaching danger. In the case of the cleft-back antelope the relationship has become more intimate and the animal's back has ceased to be a mere perch and has become a nesting site. Along the animal's back is a pair of ridges, supported by outgrowths from the vetebrae. Between the ridges is a deep cleft lined by stiff hairs that provide an ideal nesting medium for the tick bird. Several families may nest on its back at one time. Superficial warts on the animal's flanks produce pus at certain times of the year. The pus attracts flies, which lay eggs in the warts. The flies' maggots appear just as the young birds are hatching and provide them with a ready-made source of food. In return the antelope is supplied with both a constant grooming service and an early-warning system that alerts it to approaching predators.

The valuphant is a valuable element of the Lemurian ecology. In digging for the roots on which it feeds it disturbs the soil and stimulates the regrowth of vegetation.

The long-necked yippa, dependent on trees for food, migrates to the edge of the tropical forest during the dry season.

The snorke is a grass-grazer. As the herds move across the plains feeding, they expose the lower layers of vegetation, which provide food for the smaller herbivores.

# CLEFT-BACK ANTELOPE
*Castratragus grandiceros*

The tick bird's egg is held securely by stiff hairs growing in the spinal cleft.

Several families of tick bird can nest at the same time within the double ridge running down the cleft-back antelope's spine.

The cleft-back antelope is a relict of the past. It is one of the few surviving members of the large number of ungulate species that were found throughout the Age of Man.

Cleft-backs are ruminants. They have four stomachs through which their food passes to extract its maximum nutritional content.

# FLOOER
*Florifacies mirabila*

The flooer has glands around its mouth that produce a sweet-smelling secretion that is attractive to insects.

# NIGHT STALKER
*Manambulus perhorridus*

The night stalker's powerful front legs are developed from the wings of its ancestors. Its back feet, which were originally used for grasping and clutching, now come over its shoulders and effectively form hands.

# THE ISLANDS OF BATAVIA

*An island world of bats*

The streamlined aquatic surfbat is descended from a conventional flying bat ancestor. Its flippers, formed from what were once wings, have become stubby and muscular.

On land the surfbat leaps along on its tail and forelimbs. When resting its tail is curled under its body.

Although volcanic mountains and islands usually form where two crustal plates meet and crush against one another they also form over 'hot spots' on the earth's crust – areas lying above intense activity deep in the earth's mantle. Directly over the hot spot a volcano is formed. When the crust passes away from the centre of activity the volcano becomes extinct and a fresh one then erupts alongside it, producing in time a chain of progressively older volcanic islands in the middle of the ocean. During the Age of Man, a hot spot was responsible for producing the Hawaiian island chain, and in the Pacific at the present time a hot spot is in the process of generating the Batavian Islands.

Birds are usually the first vertebrates to reach and settle on new islands, but in the case of Batavia the first vertebrates to arrive were their mammalian equivalents, the bats. By the time that the birds did arrive, the bats were so well established that there were few unoccupied evolutionary niches left and the birds have never colonized the islands to any extent. The presence of suitable food on the ground, and the absence of predators enabled many bats to take up a terrestrial existence and to fill a large number of ecological niches.

The flooer, *Florifacies mirabila*, has remained an insect-eater, but is now largely sedentary. Its brightly coloured ears and nose flaps mimic a species of flower found on the islands. It sits among them with its face turned upwards, snapping at any insect that attempts to land. Although it has arisen independently, the flooer's feeding mode is remarkably similar to that of the flower-face potoo, *Griseonycta rostriflora*, of South America and is an interesting instance of convergent evolution.

The flightless shalloth, *Arboverspertilio apteryx*, is an omnivorous tree-dwelling bat which spends its life hanging upside down like the ancient sloth. It eats leaves and the occasional insect or small vertebrate caught by a swift jab of its single claw.

The beaches are home for the packs of surfbats, *Remala madipella*, which fish in the shallow waters around the coral reefs. Their hind legs, wings and tail flaps have developed into swimming and steering organs and their bodies have become sleek and streamlined. Their evolution from a flying, through a terrestrial form, into an aquatic creature is very similar to the evolutionary development of the penguin.

Once other vertebrates had established themselves on the islands, a family of ground-dwelling predator bats arose. These creatures walk on their front legs – on what would, in the case of a flying bat, be its wings, the site of most of its locomotor muscles. Their hind legs and feet are still used for grasping, but now fall forward to hang down below their chin. As the bats locate their prey purely by echolocation, their ears and nose flaps have developed at the expense of their eyes, which are now atrophied.

The largest and most fearsome of these creatures is the night stalker, *Manambulus perhorridus*. One and a half metres tall, it roams screeching and screaming through the Batavian forest at night in packs. They prey indiscriminately on mammals and reptiles, attacking them with their ferocious teeth and claws.

Excepting for the 'thumb' the fingers of the shalloth's hand-like front feet are fused together.

# THE ISLANDS OF PACAUS

*The evolution and versatility of the Pacauan whistlers*

Nut-eater
(*Insulornis macrorhyncha*)
– heavy bill for cracking shells

Insect-eater
(*Insulornis piciforma*)
– strong, pointed bill for penetrating the bark of trees

Predator
(*Insulornis aviphaga*)
– powerful hooked bill for tearing flesh

The terratail gains protection from the strong resemblance that the tip of its tail bears to the head of a bird snake.

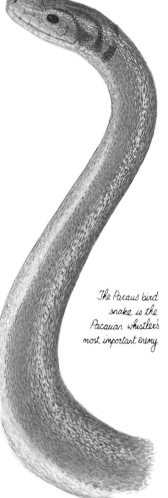

The Pacaus bird snake is the Pacauan whistler's most important enemy

Several thousand kilometres east of the Australian sub-continent lies the island chain of Pacaus. It was formed during the last 40 million years by friction between the northward-moving Australian tectonic plate and the westward-moving Pacific plate. At the margin between the two plates, volcanic islands were thrown up which gradually acquired accretions of coral round their shores.

After the ash and lava slopes were covered with vegetation and an insect population had been established, the island began to be colonized by birds. The first birds to arrive were the golden whistlers, *Pachycephala pectoralis*, which were blown across the ocean from Australia. Originally a fairly unspecialized bird it had, during the Age of Man, begun to show some differentiation, with distinct beak shapes developing on the islands around the Australian coast. However, it was only on the Pacaus Archipelago, where all the ecological niches were thrown open to them, that the whistlers really developed spectacularly, producing both insectivorous and seed-eating as well as predatory forms.

The descendants of the particular group of golden whistlers that colonized these islands are now regarded as belonging to a single genus, *Insulornis*. All the species within this genus are now highly specialized and quite different from one another excepting *I. harti*, which is similar in form to the original ancestral bird.

*I. piciforma* has developed a strong, chisel-like bill with which it tears into the bark of trees to get at burrowing insects. Its feet are modified to allow it to cling to the vertical trunks and the bird closely resembles the extinct woodpeckers of the Northern Continent whose mode of life it closely follows in almost every respect.

Nuts and tough seeds are eaten by *I. macrorhyncha*, a parrot-like species which has developed a massive bill and the powerful musculature to operate it. This bird has retained the perching feet of its ancestor and has grown a long tail to balance the weight of its large head.

All the Pacauan whistlers are preyed upon by their hawk-like relative *I. aviphaga*, which shows the same adaptations that are found in birds of prey throughout the world, irrespective of their ancestry – a hooked beak, binocular vision through forward-facing eyes, and a high degree of man-oeuvrability in pursuit.

Apart from the hawk whistler the only natural enemies faced by the Pacauan whistlers are the snakes, which have been rafted to Pacaus from Australia or the other islands in that corner of the Pacific at one time or another. The Pacauan whistlers' wariness of snakes is exploited by the terratail, *Ophiocaudatus insulatus*, a timid rodent and one of the few mammals living on the island chain. The markings on its tail mimic to a remarkable degree the markings on the head of the Pacaus bird snake, *Avanguis pacausus*, one of the most active and vicious snakes of the archipelago. When threatened by a bird, or indeed by any other creature, the terratail throws its tail into the typical snake-threat posture and utters a realistic hiss. It makes its escape rapidly into the undergrowth while its enemy is still recovering from the shock.

**PACALIAN WHISTLERS**
*Insulornis* spp.

The hawk whistler (Insulornis aviphaga), is deterred by the terratail's snake-like tail.

**TERRATAIL**
*Ophicaudatus insulatus*

When threatened the terratail thrusts out its tail and swivels its body under the branch to a position where it cannot be seen.

Apart from
isolated environments such
as caves and rocky islets, about which it is
impossible to generalize, all the earth's
major terrestrial habitats have been described.
The vast area and volume of the oceans
however have been dealt with only in passing, for although
evolution has indeed taken place in the oceans since
the Age of Man, the effects are not spectacular and would be of interest
only to the specialist.
The habitats described are not nearly so clear-cut as they might at
first seem to be. They blend into one another and certain
animals, especially the less specialized ones,
roam from one to another and cannot be truly said to belong to any
particular environment.
Any review of life at a particular period in time
gives only a two-dimensional cross-
section of a three-dimensional dynamic system which is
constantly changing and continually being
modified. Evolution is an on-going process in
which new creatures are gradually
appearing all the time while
others are becoming extinct.

# FUTURE

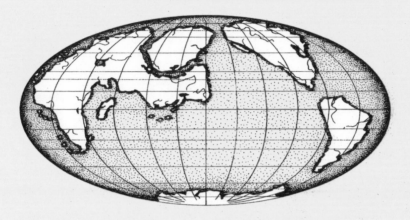

*The geography of the world one hundred million years after the Age of Man is difficult to predict, but with a knowledge of plate tectonics it is possible to suggest a distribution of land and sea that is more likely than some of the many possible patterns.*

Life will continue on the earth for as long as the earth remains in existence, which will probably be for the next 5000 million years. How life will evolve over that period there is no way of knowing, but there is one thing of which we can be sure and that is that the animals and plants will not remain as they are. The epoch following the one described on the previous pages will be characterized by a continuing movement of the earth's crust. The Atlantic Ocean may reach its maximum width within the next few tens of millions of years and begin to contract once more, bringing North America and South America back towards the European and African sub-continents. This may give rise to deep ocean troughs and new fold mountain ranges along the western side of the Northern Continent and the reopening of the Bering Strait. The result would be the isolation of North America once more and the development of new animals on that continent. It is just as likely that, within the same period of time, new convection currents may arise deep in the mantle beneath the vast Northern Continent, and a new rift system may appear on the continental mass. Such a rift system may follow one of the old sutures that indicate where earlier continents fused to form the supercontinent – such as the line of the old Ural Mountains or the Himalayan Uplands to the north of the Indian peninsula – or it may split the continent apart along a totally new line. Australia may continue to move northwards, sliding up the eastern edge of the Northern Continent, and may even tear away from it completely, isolating its fauna once more. Antarctica may, at a much later stage, also drift away from its long-established polar position. Moving into more temperate climatic belts, it would offer itself as a vast virgin continent to be settled and colonized in the normal manner.

# THE DESTINY OF LIFE

The far-reaching biological changes that will inevitably take place in the distant future will be heralded by a change in the evolution of the plants. As we have seen plants tend to evolve at a much slower rate than animals, but when a new advance does take place it has the most profound effect on animal life. The emergence of plants on to land first enabled the animals to leave the sea and to colonize the continents. The emergence of the flowering plants led to the evolution of the social insects. The extinction of the dominant tree ferns and cycads, and their replacement by broadleaved trees, led to the extinction of the great reptile groups and allowed the mammals to flourish.

It is certain that the next step in the evolution of the world's flora will lead to another such revolution in the development of animal life. Such a step is unlikely to be simple or obvious and so its prediction is something of an impossibility. It will, however, involve an increase in the efficiency of the plant's reproduction system. If that involves the replacement of seeds and fruits by another structure it will inevitably lead to the extinction of many creatures such as the birds and rodents that rely upon them for their staple diet. Other creatures will evolve in their place that will be able to reach and eat the new structures and new symbiotic relationships will develop in which the reproductive structures, in return for providing food for the animals, will be effectively fertilized or distributed by them in a way analogous to that in which birds distribute the seed of the berries on which they feed by passing them through their digestive system.

Whole new animal groups will appear independently of the floral evolution. Such groups will also rely on more sophisticated reproductive systems to give them the edge over the other groups still in existence. Further developments in sensory systems may be possible, giving an animal more awareness of its surroundings. An increase in intelligence to interpret this enhanced, sensual information would also be necessary and it may be that an intelligence as high as man's may evolve once more. Such developments may take place among the less specialized members of the most advanced groups that are around at the moment, such as the insectivores in the case of mammals or the crows in the case of birds, or they may arise from something that is with us at the moment but is so insignificant that it is constantly overlooked – after all, the mammals were scurrying about the feet of dinosaurs for some 100 million years before they came to anything. In any case many of the major groups of today will continue to soldier on even though their prominent position is usurped by newcomers; the reptiles are still around even though their day of glory has passed.

Large-scale biological disasters such as occurred at the end of the Age of Man may take place again. If that happens there will once more be the wholesale destruction of large groups of animals followed by their rapid replacement by creatures evolved from the

*In the past the evolution of new forms of animal life has corresponded with the development of new plant life; the appearance of plants on land pre-empted the first land animals (A); the social insects evolved at the same time as the flowering plants (B); the dinosaurs existed when the earth was forested by giant ferns and cycads (C) and were replaced by the mammals only when the first broad-leaved trees appeared (D). Therefore the future evolution of new animal forms will most probably coincide with developments in the plant kingdom. A likely development is one that would reduce the number of seeds that a plant must produce to provide a single offspring. In this illustration (E) the tree's seeds, although fertilized in the usual way, do not drop to the ground when ripe but remain with the parent, where they begin to grow. The plants take root and reach maturity only when they are removed by herbivores and placed in direct sunlight.*

The earth has existed in its present form for around 5000 million years and will exist for about the same length of time into the future. Life of some sort first appeared between 1000 and 1500 million years after the earth was formed and will probably continue until shortly before the earth is destroyed. The future of life is obscure. Only by looking at the events of the past can some inference be made about how life will evolve during the next 5000 million years.

A. 0 million years
*The creation of the solar system.*

B. 1250 million years
*The beginning of life – the first living forms developed.*

C. 4430 million years
*The start of the fossil record – the earliest hard-shelled creatures evolve in the world's oceans.*

D. 4570 million years
*Life begins on land – the fishes and the first land animals appear.*

E. 4720 million years
*The Age of Reptiles dawns – the earliest land-living vertebrates begin to develop.*

F. 4935 million years
*The Age of Mammals dawns – the mammals replace the reptiles as the dominant life form.*

G. 5000 million years
*The Age of Man – intelligent life appears briefly on earth. Industry and agriculture have a dramatic effect.*

H. 5050 million years
*Life after man – the earth is populated by animals that have evolved free of man's interference.*

J. 10,000 million years
*The earth destroyed – the sun begins to expand into a red giant and engulfs the inner planets.*

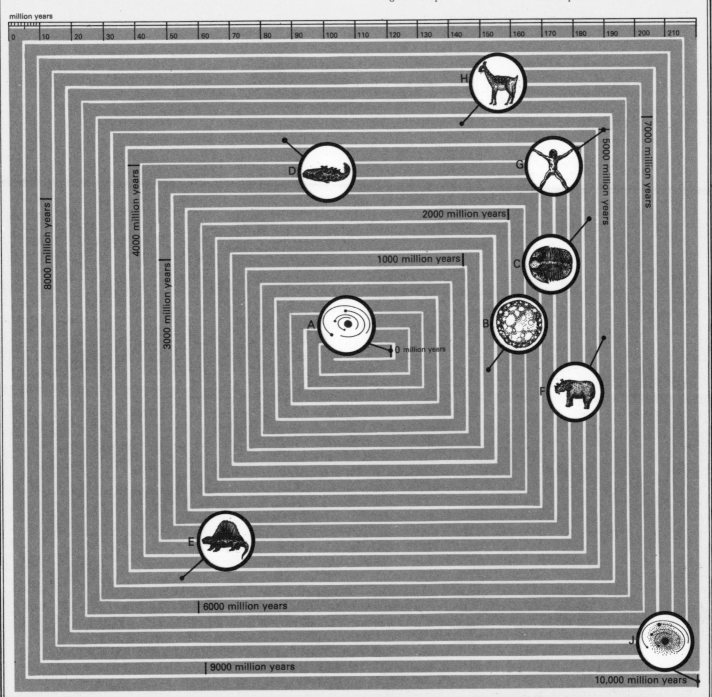

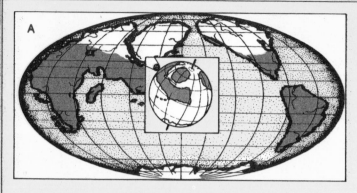

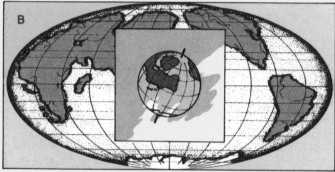

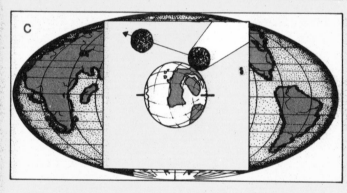

*A scale of physical events of increasing magnitude (and of increasing improbability) can be imagined, each affecting the earth's climate and therefore its animal and plant life to varying degrees. An event, such as a major meteoritic bombardment (A), which would hurl clouds of dust into the atmosphere, blocking out the sun's rays, would cause the ice-caps to advance temporarily. A more serious state of affairs would arise if the sun's rays were blocked out over several millennia as might happen if the earth were shrouded in inter-planetary dust (B) – the ice-caps would probably advance to cover most of the globe. In the extremely unlikelihood of a cosmic collision large enough to disturb the earth's orbital and axial alignment (C), the effect would be permanent catastrophic and impossible to predict.*

could disrupt the composition of the atmosphere would undoubtedly cause widespread extinctions and may even render the continents totally barren. However, no matter how great the atmospheric damage is, there will still be some organism somewhere that will survive, even if such an organism is nothing more than a simple cell – the nature of life, as we have seen, is such that it is able to replicate itself and fill all possible niches. Evolution will begin again, the seas will teem with life once more and eventually the land will be recolonized.

What this new phase of earth's evolution will look like is impossible to predict, but we can be sure that the new animals will look nothing like those that we have known. The possibilities of genetic development are infinite and the surviving systems could be selected from innumerable possible combinations. Convergent evolution will not be able to reinstate the kinds of animals and plants with which we are familiar since the basic evolutionary stock will be so different and the niches to be occupied will be nothing like those we know now. An even bigger meteorite could destroy even the crustal fabric of the earth, but the larger the disaster we postulate the more unlikely it is to happen during the next 5000 million years.

By that time the sun will have used up all of the hydrogen available to it. Its core will have shrunk and its surface will have become much cooler. The sun's helium will then begin to react and the sun to expand, increasing its luminosity by many hundreds of times. This will mark the end of life on earth. As the temperature rises the organic reactions that support life will be no longer possible. The seas will boil away and the atmosphere will be stripped off. As the sun, now a red giant, continues to expand it will engulf all of the inner planets, including the earth. Before long all the material capable of supporting nuclear energy in the sun will be used up and, very rapidly, in terms of the geological time scale, it will collapse into a fraction of its former volume. The gravitational force involved in this collapse will make it shine as a white dwarf until, with all its energy dissipated, it fades into a dark cold lump – a black dwarf. The planets, if they physically survive, will be no more than dark cinders devoid of water and atmosphere and, incapable of supporting life ever again.

However, the chemical and physical reactions that took place to produce life on earth will take place, or may very well have taken place, again on other planets, in other solar systems. Such life forms will be specifically evolved to match the conditions found there, although what these conditions will be, and hence the forms of life that will evolve to cope with them, cannot possibly be imagined. It is almost certain that life will always exist somewhere in the universe in one form or another.

survivors. Despite the very great temporary change to the environment and ecosystem, such a disaster is unlikely to have a long-term detrimental effect on life as a whole.

Physical, non-biological disasters are possible, such as the impact of a large meteorite with the earth. If such a meteorite were large enough the resulting explosion may release vast volumes of dust into the atmosphere and reduce the intensity of solar radiation at the earth's surface quite considerably for a number of years. The result would be a decrease in plant growth, with the accompanying reduction in herbivorous animals and drastic effects on the populations of the carnivores.

An increase of the volume of interplanetary dust in the solar system will have a similar effect in reducing the amount of sunlight reaching the earth. The climatic effects of such occurrences will be far-reaching. The temperature of the earth's surface will fall and the ice-caps of the poles will grow and reach towards the equator. When such ice ages have happened on earth in times gone past they have led to the evolution of animals and plants equipped to stand up to such rigours rather than to any great damage to the basic structure of life itself.

The impact of a body huge enough to produce shock waves that

# APPENDIX
## GLOSSARY

**Adaptive radiation** The expansion of a single species into a number of new forms that are capable of filling a variety of vacant ecological niches. *See* the bats of Batavia (p.109).

**Allen's rule** In animal groups with a large north–south range, the species or subspecies nearer the poles have smaller extremities. *See* the temperate ravene (p.40) and the polar ravene (p.63).

**Batesian mimicry** The resemblance of a harmless species to a dangerous or distasteful one to gain protection by association, *c.f.* Müllerian mimicry. *See* the terratail (p.110).

**Bergmann's rule** In animal groups with a large north–south range, the species or subspecies nearer the poles will be larger. *See* the rabbucks (p.38).

**Brachiation** Swinging by the hands and arms, a means of locomotion typical of tree-dwelling primates. *See* the ziddah (p.88).

**Brood Parasite** A creature that leaves its offspring with the brood of another to be cared for and tended by the parents of that brood. *See* the gandimot (p.63).

**Browser** An animal that eats leaves and shoots, *c.f.* Grazer. *See* the zarander (p.92).

**Carnivore** In general terms, an animal, either a predator or a scavenger, that eats meat. More precisely the term is restricted to members of the order Carnivora. *See* the pamthret (p.54).

**Cline** A chain of subspecies. *See* the flightless auks (p.64).

**Commensalism** Living together or sharing the same food supply to gain mutual benefit. This relationship is not essential for the survival of any of the parties involved, *c.f.* Parasitism *and* Symbiosis. *See* the meaching and the lesser ptarmigan (p.63).

**Convergent evolution** The evolution of similar physiological or anatomical features by unrelated groups of animals. *See* the flooer (p.109) and the flower-faced potoo (p.105).

**Countershading** A pattern of coloration in which the upper side of an animal is darker than the lower side. Countershading destroys the natural pattern of light and shade and makes an animal inconspicuous. *See* the picktooth (p.81).

**Dentition** The number, type and pattern of teeth. Fish, amphibians and reptiles have teeth that are all of the same shape and size. Mammals have teeth of various types: incisors (front teeth) for cutting and grasping, canines for piercing and premolars and molars (back teeth) for grinding and shearing.

**Dimorphism, sexual** A marked difference in structure or appearance between the sexes of the same species. *See* the matriach tinamou (p.102).

**Ecological niche** The station occupied by an animal in an environment. The ecological niche determines a creature's mode of life, e.g. tree-dweller, grazer, etc.

**Evolution** The development of new species from earlier forms. Evolution can also apply to the development of the features of an animal's anatomy. These can be referred to as:

> **Analogous** Features similar in form or function to one another but which have evolved from different structures. *See* tails of the mud-gulper (p.95) and distarterops (p.64).
>
> **Homologous** Features which may be different in function or appearance but have the same origin as one another. *See* paddles of the surfbat (p.109) and walking limbs of the night stalker (p.109).
>
> **Degenerate** A feature or an organism of less sophistication than its predecessors. *See* dentition of the turmi (p.92).
>
> **Primitive** An unsophisticated feature that has remained with a creature throughout its evolutionary history. *See* tail of the falanx (p.40).
>
> **Secondary** A feature which was once possessed by an ancestor and subsequently lost to be redeveloped at a later evolutionary stage. *See* numbers of neck vertebrae in the reed stilt (p.49).

**Genus** The biological classification that embraces a group of related species.

**Grazer** An animal that eats grass, *c.f.* Browser.

**Insectivore** An animal that eats insects. More precisely the term is restricted to members of the order Insectivora. *See* the pfrit (p.49).

**Invertebrate** An animal without a backbone.

**Marsupial** A mammal giving birth to its young in an undeveloped state. Most marsupial young are reared in pouches situated on their mothers' abdomens, *c.f.* Placental. *See* the chuckaboo (p.97).

**Müllerian mimicry** The resemblance between a group of dangerous or distasteful species to gain mutual benefit, *c.f.* Batesian mimicry *See* desert predators (p.75).

**Natural Selection** The persistence of animals best able to survive.

**Opposability (fingers)** The ability of one fingertip to touch the other fingers on the same hand. *See* the striger (p.91).

**Omnivore** An animal that eats both plant and animal material.

**Parallel evolution** The evolution of similar anatomical or physiological features by related groups of animals. *See* the gigantelope (p.82) and the valuphant (p.106).

**Parasitism** The feeding of one organism directly on another without the host gaining any benefit in return, *c.f.* Commensalism *and* Symbiosis. *See* the trovamp (p.92).

**Patagia** Flaps of skin used as wings in gliding animals. *See* the flunkey (p.88).

**Photosynthesis** The transformation within plants of inorganic nutrients into food using sunlight.

**Placental** A mammal that possesses a placenta through which it feeds its embryo young in the womb, *c.f.* Marsupial.

**Plankton** Animals and plants, mostly microscopic, that float passively in the water.

**Plate tectonics** The study of the major geological plates that make up the earth's crust.

**Predator** An animal that actively hunts, kills and eats other animals.

**Prehensile** Able to grasp and hold things – a term normally applied to tails. *See* the ziddah (p.88).

**Primates** The order of mammals that includes the apes and monkeys.

**Prosimians** The group of primates that includes lemurs and lorises. *See* the clatta (p.91).

**Scavenger** An animal that eats the dead bodies of others. *See* the ghole (p.84).

**Symbiosis** Living closely together for reasons of mutual survival, *c.f.* Commensalism *and* Parasitism. *See* the cleft-back antelope and the tick bird (p.106).

**Ungulate** In general terms an animal with hooves, more precisely a member of the orders Artiodactyla and Perissodactyla. *See* the island continent of Lemuria (p.106).

**Vertebrate** An animal possessing a backbone.

**Wattle** A fleshy appendage on the throat of a bird. *See* the flightless guinea fowl (p.81).

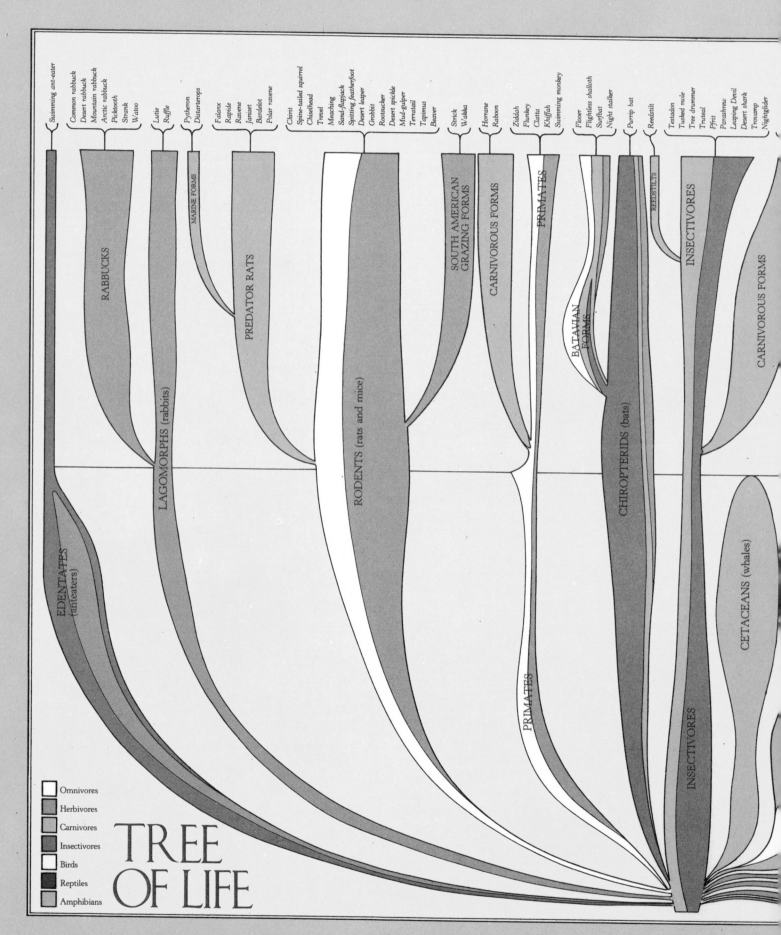

Swimming ant-eater

Common rabbuck
Desert rabbuck
Mountain rabbuck
Arctic rabbuck
Picktooth
Strank
Watoo

Lutie
Ruffle

Pytheron
Distarterops

Falanx
Rapide
Ravene
Jamset
Bardelot
Polar ravene

Chirit
Spine-tailed squirrel
Chiselhead
Trewel
Meaching
Sand-flapjack
Spitting featherfoot
Desert leaper
Grobbit
Rootsucker
Desert spickle
Mud-gulper
Terratail
Tapimus
Beaver

Strick
Wakka

Horrane
Raboon

Ziddah
Flunkey
Clatta
Khiffah
Swimming monkey

Flooer
Flightless shalloth
Surfbat
Night stalker

Pump bat

Reedstilt

Testadon
Tusked mole
Tree drummer
Truteal
Pfrit
Parashrew
Leaping Devil
Desert shark
Trovamp
Nightglider

RABBUCKS

MARINE FORMS

LAGOMORPHS (rabbits)

PREDATOR RATS

SOUTH AMERICAN
GRAZING FORMS

CARNIVOROUS FORMS

PRIMATES

BATAVIAN FORMS

RODENTS (rats and mice)

CHIROPTERIDS (bats)

REEDSTILTS

INSECTIVORES

CARNIVOROUS FORMS

EDENTATES
(anteaters)

PRIMATES

INSECTIVORES

CETACEANS (whales)

Omnivores
Herbivores
Carnivores
Insectivores
Birds
Reptiles
Amphibians

TREE
OF LIFE

118

Singet
Gurrath
Zarander
Turmi
Valluphant
Snorke
Cleft-back antelope
Helmeted hornhead
Common hornhead
Water hornhead
Woolly gigantelope
Groath
Tropical gigantelope
Long-necked gigantelope
Rundihorn
Chuckaboo
Slobber
Hiri-hiri
Posset
Giantala
Vortex
Porpin
Angler heron
Flightless auk
Long-legged quail
Toothed kingfisher
Shern
Pilofile
Lesser ptarmigan
Giant pitta
Hawkbower
Termite burrower
Tick bird
Flower-faced potoo
Flightless guinea fowl
Tree duck
Long-necked dipper
Bootie bird
Gandimot
Hanging bird
Broadbeak
Common pine chuck
Pacauan whistlers
Matriach tinamou
Fin lizard
Anchoruhip
Fatsnake
Pacaus bird snake
Oakleaf toad

CARNIVORES

BROWSING FORMS

ANTEATER FORMS

PRIMITIVE FORMS (Lemuria)

GIGANTELOPES AND HORNHEADS

MARSUPIALS

PELAGORNIDS

ARTIODACTYLS (pigs)

ARTIODACTYLS (ruminants)

PERISSODACTYLS
(horses and rhinos)

PRIMITIVE SOUTH AMERICAN
UNGULATES

PROBOSCIDEANS

MARSUPIALS

MONOTREMES

"ADVANCED" NON-PERCHING BIRDS

"ADVANCED" PERCHING BIRDS

Crocodiles

Lizards and snakes

Turtles

Frogs and toads

Newts and salamanders

"PRIMITIVE" BIRDS

Crocodiles

Lizards and snakes

Turtles

Frogs and toads

Newts and salamanders

50 million years
after man.

Australia links with
Asia.

Mediterranean Sea
closes –
South America splits
off from North America
– the Bering Straits
close.

Age of Man.

Pleistocene ice-ages.

South America unites
with
North America.

Indian sub-continent
becomes part of Asia.

Global climate
becomes drier.
Grasslands appear.

65 million years
before man.

50

40

30

20

10

0

10

20

30

40

50

60

65

# INDEX

Page numbers in *italic* refer to the illustrations and their captions.

# ACKNOWLEDGEMENTS

The author would like to thank Malcolm Hart for his help in predicting the bird life found on earth in fifty million years' time and Dr. John Oats for his advice and criticism in the preparation of the text.

Index prepared by Hilary Bird

## BIBLIOGRAPHY

Bourlière, F., *The Natural History of Mammals*, George G. Harrap (London, 1955).

Cloudsley-Thompson, J.L., *Terrestrial Environments*, Croom Helm (London, 1973).

Colinvaux, P., *Why Big Fierce Animals Are Rare*, Allen & Unwin (London, 1978).

Dawkins, R., *The Selfish Gene*, Granada (London, 1978).

Gotch, A.F., *Mammals – Their Latin Names Explained*, Blandford (Poole, 1979).

Gould, S.J., What's Wrong With Marsupials? *New Scientist*, Vol. 88, No. 1221 (1980).

Halstead, L.B., *The Pattern of Vertebrate Evolution*, Oliver and Boyd (Edinburgh, 1969).

Hoyle, F. & Wickramasingbe, N.C., *Lifecloud, The Origin of Life in the Universe*, J.M. Dent (London, 1978).

Koob, D.D. & Boggs, W.E., *The Nature of Life*, Addison-Wesley (Reading, Massachusetts, 1972).

Kurtén, B., Continental Drift and Evolution, *Scientific American*, (March, 1969).

Lawrence, M.L. and Brown, R.W., *Mammals of Britain, Their Tracks, Trails and Signs*, Blandford (Poole, 1974).

Mitchell, J. (ed.) *The Natural World* volume of *The Mitchell Beazley Joy of Knowledge Library*, Mitchell Beazley (London, 1977).

Perry, R., *Life in Forest and Jungle*, David & Charles (Newton Abbot, 1976).

Rostrand, J., *Evolution*, Prentice Hall (London, 1960).

Simon, S. & Bonners, S., *Life on Ice*, Watts (London, 1976).

Simpson, G.G., *Splendid Isolation, The Curious History of South American Mammals*, Yale University Press (New Haven and London, 1980).

Stebbins, G.L., *Processes of Organic Evolution*, Prentice-Hall (New Jersey, 1977).

Young, J.Z., *The Life of Vertebrates*, University Press (Oxford, 1962).

## SOURCES OF REFERENCE

Beerbower, J.R., *Search for the past*, Prentice-Hall (Englewood Cliffs, N.J., 1968)

Benes, J., *Prehistoric Plants and Animals*, Hamlyn (London, 1979)

Bramwell, M. (ed.), *The World Atlas of Birds*, Mitchell Beazley (London, 1974)

Carthy, J.D., *The Study of Behaviour*, Edward Arnold (London, 1979)

Clark, D.L., *Fossils, Palaeontology and Evolution*, Wm. C. Brown (Dubuque, Iowa, 1968)

Dietz R.S. & Holden J.C., The Brakeup of Pangaea, *Scientific American* (October 1970)

Fenton & Fenton, *In Prehistoric Seas*, George Harrap (London 1964)

Gillie, O., *The Living Cell*, Thames & Hudson (London, 1971)

Mackean, D.G., *Introduction to Genetics*, John Murray (London, 1977)

Moore, R., *Evolution*, Time-Life (London, 1973)

Pfeiffer, J., *The Cell*, Time-Life (London, 1972)

Phillipson, J., *Evolutionary Energetics*, Edward Arnold (London, 1966)

Romer, A.S., *The Vertebrate Story*, University of Chicago (Chicago, 1959)

Scott, J., *Palaeontology*, Kahn & Averill (London, 1973)

Spinar, Z.V., *Life before Man*, Thames and Hudson (London, 1972)

Swinnerton, H.H., *Outlines of Palaeontology*, Edward Arnold (London, 1947)

Whitfield, P. (ed.), *The Animal Family*, Hamlyn (London, 1979)

The illustration on page 24 was redrawn from the Cambridge Bible of 1663.

## ILLUSTRATORS

Diz Wallis, pages: 38–39; 40–41; 42; 44–45; 46; 48–49; 52; 54–55; 56–57; 60–61; 63; 64; 66–67; 68; 72–73; 75; 76–77; 80–81; 82; 84–85; 88–89; 90–91; 92–93; 95; 96–97; 98; 102–103; 105; 106; 108–109; 110–111

John Butler, pages: 43; 47; 65; 69; 83; 94; 104

Brian McIntyre, pages: 36; 50; 58; 70; 78; 86; 100; 112

Philip Hood, pages: 53; 62; 74; 99; 107

Roy Woodard, pages: 23; 33; 34–35; 37; 51; 59; 71; 79; 87; 107; 113

Gary Marsh, pages: 11; 12–13; 14–15; 16–17; 18–19; 20–21; 22; 24–25; 26–27; 28–29; 30–31; 32; 114–115; 116; 118–119